Simeon Hayden Guilford

Orthodontia, or Malposition of the Human Teeth

Its prevention and remedy. Second Edition

Simeon Hayden Guilford

Orthodontia, or Malposition of the Human Teeth
Its prevention and remedy. Second Edition

ISBN/EAN: 9783337088972

Printed in Europe, USA, Canada, Australia, Japan

Cover: Foto ©berggeist007 / pixelio.de

More available books at **www.hansebooks.com**

ORTHODONTIA,

OR

MALPOSITION OF THE HUMAN TEETH;

ITS PREVENTION AND REMEDY.

BY

S. H. GUILFORD, A.M., D.D.S., Ph.D.,

PROFESSOR OF OPERATIVE AND PROSTHETIC DENTISTRY IN THE PHILADELPHIA DENTAL COLLEGE; AUTHOR OF "NITROUS OXIDE," &C.

Approved by the National Association of Dental Faculties as a text book for use in the schools of its representation.

SECOND EDITION, REVISED AND ENLARGED.

PHILADELPHIA:
PRESS OF SPANGLER & DAVIS,
529 COMMERCE STREET.

Entered according to Act of Congress, in the year 1893, by
S. H. GUILFORD,
In the Office of the Librarian of Congress, at Washington, D. C.

TO MY
FELLOW-TEACHERS
OF THIS SPECIAL BRANCH OF
DENTAL SCIENCE AND TO THE MANY
PUPILS WHOM IT HAS BEEN MY PLEASURE TO
INSTRUCT, THIS SECOND EDITION IS RESPECTFULLY INSCRIBED.

Preface to First Edition.

This work has been written at the request of the National Association of Dental Faculties in furtherance of its plan to secure the preparation of a series of text-books for use in American Dental Colleges. After its completion and examination, it was accepted and endorsed by the Association at its meeting in Saratoga, August, 1889.

The impartment of instruction in the simplest and most direct manner being the true province of a text-book, the author has endeavored in the preparation of this work to treat the subject as concisely as possible, and to clothe his thoughts and those of others in such language as to be readily comprehended by beginners as well as those somewhat advanced in this branch of study.

In the treatment of the subject, the aim has been to lead the student step by step from the simplest beginnings to the more complicated and difficult work of practical treatment. To this end, the underlying principles of the art are first elucidated, after which the principal methods employed are explained, and lastly, the correlation of principles and methods is shown in their practical application to typical cases. In Part III, the different forms of irregularity, together with a variety of plans for their correction, are arranged under such headings and in such order as to be readily referred to in seeking aid for cases that occur in office practice.

Should the work fulfill the object aimed at in its preparation, the author will feel amply repaid.

Credit for assistance is most cheerfully given to the twenty-five teachers of this branch in American Dental Colleges who have read this work in manuscript, and by friendly criticism and valuable suggestions added much to its completeness.

The author would also acknowledge his indebtedness to Prof. W. F. Litch for valuable services, and to the S. S. White Co.; Lea, Brothers & Co.; P. Blakiston, Son & Co.; and other publishers and authors for the use of certain cuts.

S. H. G.

Philadelphia, Sept., 1889.

Preface to Second Edition.

The exhaustion of the first edition of this work within three years, conjoined with assurances of appreciation received from teachers of this branch in many dental colleges, and its selection for translation into French, and publication first in serial and later in book form by the editor of *Le Progrès Dentaire*, leads the author to believe that the work has fulfilled its mission and been of service to those engaged in this line of study.

In the preparation of the present edition it has been the aim of the author to bring it fully up to the present state of knowledge in this rapidly-advancing specialty, to do which it has been found necessary to rewrite almost the entire work and enlarge it by more than forty pages. Each chapter has been emended so as to exclude such methods of treatment as were found to be of lesser value and to include in their stead others that are newer and of greater practical importance. Some thirty of the former illustrations have been discarded, and more than fifty new ones introduced to make clearer the meaning of the text.

Two new chapters have been added; one on the "Construction of Appliances" and another on "Electro-Plating." The former, fully illustrated, is designed to instruct and assist the student in the construction of the various appliances mentioned throughout the work, and the latter, to enable him to impart to the devices made of baser metals a more sightly appearance and purer surface. These chapters, with part of Chapter X, Part III, of the former edition have been combined to form Part IV. of the present one.

The author desires to express his indebtedness to many co-laborers in this branch for valuable suggestions, and especially to Professors C. L. Goddard, of San Francisco, A. E. Matteson and C. S. Case, of Chicago, E. H. Angle, of Minneapolis, and Dr. V. E. Jackson, of New York, for the loan of models and appliances and the description of cases.

Acknowledgment is also due the S. S. White Dent. Mfg. Co., Wilmington Dent. Co., H. D. Justi & Son and other publishers for their kindness in permitting the reproduction of many illustrations.

S. H. G.

Philadelphia, March, 1893.

CONTENTS.

PART I.
PRINCIPLES INVOLVED.

CHAPTER I.
PAGE.
REGULARITY AND IRREGULARITY DEFINED,   9

CHAPTER II.
ETIOLOGY.

HEREDITY—LONG RETENTION OF DECIDUOUS TEETH—EARLY EXTRACTION OF DECIDUOUS TEETH—INJUDICIOUS EXTRACTION—DELAYED ERUPTION—SUPERNUMERARY TEETH—ACCIDENTS—ADENOID VEGETATIONS—HABITS—SUPERIOR PROTRUSION—PROGNATHISM—V-ARCH—SADDLE-ARCH,   12

CHAPTER III.
EVILS ASSOCIATED WITH IRREGULARITY.

APPEARANCE MARRED—SPEECH AFFECTED—MASTICATION IMPAIRED—CARIES INDUCED,   29

CHAPTER IV.
ADVISABILITY OF CORRECTION.

AGE—HEALTH—SEX—POWER OF APPRECIATION—FAMILY TYPE—IMPROVEMENT OF OCCLUSION,   32

Chapter V.

AGE AT WHICH CORRECTION MAY BE BEGUN.

Early Interference—When Justifiable and Advisable—When Correction should be Delayed, . 37

Chapter VI.

MOVEMENTS TO BE PRODUCED.

Principles Governing Application of Force—Rules as to same, ... 42

Chapter VII.

EXTRACTION AS RELATED TO ORTHODONTIA.

Rules Governing same, . 48

Chapter VIII.

PHYSIOLOGY OF TOOTH-MOVEMENT.

Character of Tissues Involved—The Alveolar Process—The Teeth—The Pulp—The Pericementum—Physiological Action in Movement of Teeth, . 56

PART II.

MATERIALS AND METHODS.

Chapter I.

Examination of Mouth—Impression and Articulation—Study of Case from Articulated Models, . 64

Chapter II.

APPLIANCES.

Materials and their Uses—Qualities an Appliance should Possess—Retaining Appliances, . . 75

Chapter III.

CONSIDERATION OF METHODS.

Farrar's—Patrick's—Byrnes'—Magill Band—Angle's—Coffin's—Jackson's, . . . 90

PART III.

SPECIFIC FORMS OF IRREGULARITY AND THEIR TREATMENT, ... 117

Chap.	I.	Incisor Teeth Erupting Outside or Inside of Arch. Methods of Prevention and Correction, ...	118
"	II.	Delayed or Mal-Eruption of Permanent Cuspids, ...	125
"	III.	Incisor Teeth Situated Outside or Inside of Arch after Dentition is Complete—Cases Illustrating Condition and Treatment,	128
"	IV.	Cuspid Teeth Situated Outside or Inside of Arch—Cases and Treatment, . .	138
"	V.	Misplaced Bicuspids, ...	151
"	VI.	Torsion—Double Torsion, . . .	157
"	VII.	Contraction of Arch, . . .	167
"	VIII.	Protrusion of Upper Jaw, . . .	175
"	IX.	Protrusion of Lower Jaw or Prognathism—Allen's Device—Kingsley's—Winner's Case—Author's, ...	188
"	X.	Lack of Anterior Occlusion—Various Plans for Treatment, . . .	194

PART IV.

Chap.	I.	Crowded Lower Incisors, ...	197
"	II.	Reduction of Elongation of the Anterior Teeth, ...	201
"	III.	Assisted Eruption of the Anterior Teeth,	204
"	IV.	Tooth-Shaping, ...	208
"	V.	Construction of Regulating Appliances—Description and Illustration of Tools and Appliances—Ferrules or Bands—Round Tubing—Square Tubing—Wire Drawing—Soft Soldering—Hard Soldering—Screws and Nuts—Comparative Table of Gauges.	211
Chap.	VI.	Electro-Gilding. Cleansing Baths—Copper Solution—Gold Solution—Battery—Plating, .	223

ORTHODONTIA.

PART I. PRINCIPLES INVOLVED.

CHAPTER I.

DEFINITION OF SUBJECT.

Orthodontia, from $ορθος$, straight, and $οδους$, a tooth, is that branch of dental practice which relates to the correction of irregularity of position of the human teeth.

Its recognition as a distinct branch or specialty of general dental practice has come about in recent years, indeed, it attracted so little attention less than a century ago that many of the writers of that day entirely omitted it from their books and writings, while those who did refer to it, gave it but little attention and space. Whether the condition of irregularity was less frequently met with then than now we cannot certainly tell, but inasmuch as dentistry was then in its infancy, and as the most pressing demands upon the dentist of that day were for the alleviation of pain, the substitution of artificial dentures to replace lost members, and the checking of the ravages of decay by filling, it is but natural to suppose that there was little time or inclination to attempt the relief of so apparently unimportant a condition as mere irregularity of position. Since then, however, with the natural growth of dental science and the enlargement of its sphere, the subject of orthodontia has grown in importance until to-day it is engaging the attention of some of the best minds in the profession and forms an important part of the study of every dental student.

With the growth of its interest and importance, there has been a corresponding advance in investigation as to the cause and frequency of irregularities, a more exact microscopical examination of the tissues concerned and of the physiological changes occurring in them in the process of correcting such conditions; progress has also been marked by the invention of a multiplicity of devices and appliances for the more perfect and easy correction of this class of deformities.

REGULARITY AND IRREGULARITY DEFINED.

The teeth of man when normally placed in the alveolar arch describe in outline a parabola or semi-ellipse with a slight flattening of the curve in the region of the incisor and bicuspid teeth, and a consequent tendency to angularity where the cuspids are placed. The lower arch differs from the upper principally in being slightly smaller. The teeth when thus placed should be in contact, each one touching its neighbor at the most prominent points of its approximal surfaces, and with the cusps or occluding surfaces properly articulating with those in the opposite jaw. When thus arranged the teeth are called regular.

An irregularity may be defined as any variation from the above order. It may consist in a deviation from the normal outline on the part of several or all of the teeth, or in the malposition of one or more individual teeth; if the latter, the tooth or teeth may be found outside or inside of the regular line of the arch or they may be placed anteriorly or posteriorly to their normal positions, or finally, they may be turned or twisted on their axes. In many cases this torsion is associated with malposition.

An irregularity being an abnormality, corrective measures, as a rule, should be resorted to, but slight irregularities do not always demand interference.

The slight overlapping of the superior centrals by the laterals, for instance, is a clear case of irregularity, but it is so slight and so commonly met with, that it has almost ceased to attract attention or to be regarded as an abnormality. Artificial teeth are now made reproducing this condition and in many cases are preferred on account of their " more natural appearance."

So too, the slight irregularity commonly found in connection with the inferior incisors, where several or all of them are slightly turned and overlapping, is no longer looked upon as inharmonious and is also imitated in the arrangement of artificial teeth.

Again, the slight misplacement of a tooth in the posterior part of the arch, where it is not noticeable, may be left without disturbance and no harm result.

In cases like these, if the slightly altered position of the individual teeth is not likely to result in injury to tooth structure, or does not interfere with speech or occlusion, it is best to omit any effort toward correction.

CHAPTER II.

ETIOLOGY.

The causes responsible for the production of irregularity are many and at best but imperfectly understood. Some of them are operative before the birth of the individual and others afterward. They may therefore be classed under the two general heads of Hereditary and Acquired.

HEREDITARY.

This class comprises all such cases as are evidently due to the inheritance of peculiarities that existed in their near or remote ancestors, or to some of the characteristics of both parents who were themselves free from dental abnormality.

The well-known biological law of transmission of characteristics from parent to child will readily explain how the abnormalities as well as the normalities may be transmitted. The child may bear a close resemblance to either parent in form and feature, or it may combine some of the peculiarities of both. In other cases it will resemble neither, but be like one of the grandparents or other remote relatives.

The evidences of inheritance are perhaps nowhere more clearly expressed than in the dental organs. Not only in these organs as a whole may we see the dental apparatus of a progenitor reproduced in entirety, but the resemblance is equally well shown in the inheritance of so slight an abnormality as a turned or misplaced tooth. Sometimes such peculiarity may be inherited by several children in the same family.

Cases of irregularity due to inheritance are oftentimes the most difficult to correct, for not only must mechanical difficulties be overcome, but in addition the influence of physical impress, confirmed perhaps by repeated transmission, must be combatted. The mechanical difficulties in such cases are as readily conquered as in others, but the force of inheritance will show itself in a strong and stubborn tendency on the part of the teeth to return to their former abnormal position.

The intermarriage of races with widely differing characteristics has come to be regarded as one of the most prolific causes of dental irregularity. If both races represented in the marriage possess somewhat similar characteristics as to size, vigor and feature, no dental peculiarity will usually be found in the offspring; but where the differences are marked, irregularity of the teeth will often be the result.

When one parent possesses a large frame with full-sized teeth set in large jaws and the other a small frame with correspondingly small jaws and small teeth, the child may inherit the large teeth of one parent and the small jaws of the other. The small jaws cannot accommodate the full complement of the larger teeth, and hence a crowded and irregular dental arch will be the result.

Where the small teeth of one parent and the large jaws of the other are found united in the offspring, abnormal interdental spaces will frequently be the result. These spaces may exist between all of the teeth, or, as in some cases, the deformity will only be found in connection with the anterior ones. Cases of this character, fortunately, are infrequently met with, but when they occur they present an unsightly appearance, and generally result in an earlier loss of the teeth from that lack of contact and mutual support so necessary to their longest retention and usefulness.

ACQUIRED.

The causes productive of irregularity during dentition or subsequent to it far exceed in number those due to heredity.

LONG RETENTION OF DECIDUOUS TEETH.

In accordance with physiological law, the deciduous teeth are intended to subserve the wants of the child until replaced by the permanent set. The crown of the permanent tooth should occupy a position beneath or adjacent to the root of the deciduous one which it is intended to supplant. Then, as the root of the temporary tooth is gradually removed, the permanent tooth advances and finally occupies the position previously occupied by its predecessor.

It frequently happens, however, that the crypt of the permanent tooth is situated at some little distance from the root of its corresponding deciduous one, and as the new tooth makes its way into place it assumes a position to the side of the deciduous root. As usually that part of the root is absorbed which is in contact with the vascular covering of the advancing crown, a portion of the length of the root remains unabsorbed, and the new crown is, in consequence, compelled to advance by the side of the root instead of beneath it. The deciduous tooth as a result of its only partially absorbed root, remains firm in place, and the new one is erupted out of its proper position. Had the condition been brought to the knowledge of the dentist before the new crown appeared, the extraction of the deciduous tooth would have permitted the advancing tooth to assume its proper position in the arch and irregularity have been prevented. When the permanent tooth is advancing out of position the fact may be recognized by the unusual distension of the gum and alveolar plate beneath, and the deciduous tooth, no matter how firmly set, should at once be removed. Even the spicula of a deciduous root has been found sufficient to deflect a permanent tooth from its course during eruption.

EARLY EXTRACTION OF DECIDUOUS TEETH.

That the premature extraction of deciduous teeth often prepares the way for irregularity of the permanent set is generally recognized, but the extent of its importance and the manner in which it operates can best be understood by considering the physiological facts in the case.

Irregularity of the deciduous teeth is a condition very seldom met with. As a rule they occupy their normal positions in an alveolar arch of proper size to accommodate them, and this again rests upon a jaw bone of suitable amplitude. Thus jaw, process and teeth are harmoniously correlated. As each deciduous tooth is lost it is succeeded by the corresponding permanent one, which, under normal conditions, will occupy the space created by the removal of its predecessor. In this way, one by one, the permanent set should make its appearance until all of the deciduous teeth have been supplanted by their permanent successors.

The permanent teeth are all larger than the corresponding ones of the deciduous set, with one exception,—the second bicuspid. This being the case, they require a larger alveolar arch and a correspondingly larger jaw bone for their accommodation. This nature furnishes by the slow process of enlargement by interstitial growth, which is hastened and stimulated by the lateral pressure of the teeth as they make their way into place, and afterward. When the first permanent molar makes its appearance it is obliged to provide sufficient accommodation for itself by forcing its way between the deciduous second molar and the strong maxillary tuberosity above or the equally resistant ramus below. This pressure is felt by all the other teeth in the arch. If, therefore, any of the deciduous molars should be extracted about the fifth or sixth year, for instance, as they too often are after having been impaired by disease, the permanent molar will move forward and occupy part of the space intended for the bicuspids.

When the permanent lower central incisors erupt they make their appearance inside of the deciduous ones, which soon loosen and drop out. Owing to the fact that the width of these new teeth is considerably greater than that of their predecessors, they naturally overlap to a certain extent the adjoining deciduous laterals. This overlapping prevents the centrals from moving forward into line in the arch. When the permanent laterals erupt they assume a position by the side of the centrals, and to find accommodation in this contracted space inside of the arch several or all of them are apt to be crowded into irregular positions.

This condition, from the fact that these teeth have erupted too rapidly to admit of a corresponding increase in size of the alveolar arch, is often regarded as a serious evil, and to correct it, the inexperienced practitioner will in many cases extract the temporary cuspids which are designed for retention until years afterward. This additional space having been thus furnished, the permanent incisors will move forward into line and assume a regular position.

Later, when the bicuspids appear, they will usually find no difficulty in assuming places in the arch, because their predecessors occupied a larger space and because the cuspids are missing, but from the very abundance of the space and the pressure of the first molar from behind, the bicuspids will very soon, if not at once, be so pressed forward that the first bicuspid will come in contact with the lateral, leaving no space for the accommodation of the cuspid when it makes its appearance at about the eleventh or twelfth year.

Such being the case, the cuspid must of necessity erupt outside or inside of the arch, and produce a deformity both unsightly and hard to correct.

Had the temporary cuspids not been extracted, they would have preserved space for their successors, and the inlocked and irregular incisors, in the course of time, by the normal enlargement of the arch, and the excess provided by the removal of the deciduous molars, would have had space sufficient, which nature, assisted by the pressure of the tongue, would aid them in occupying.

The same condition is met with in the superior arch, perhaps more frequently than in the inferior. Here the incisors erupt outside of the deciduous ones, and sometimes appear in an irregular and crowded position, to correct which the temporary cuspids are often needlessly sacrificed, and the same train of evils follows.

It will thus be seen that the premature extraction of any of the temporary teeth, especially the cuspids, cannot well result in other than harm to the permanent ones, so far as regularity is concerned.

Sir John Tomes relates a case in which he extracted for cause all of the deciduous teeth of a child, and yet when the permanent ones appeared they assumed their proper positions in the arch without any resultant irregularity.

This one case, however, the only one of the kind on record, does not disprove the facts as noticed in thousands of cases of opposite character, nor does it confute the plainly apparent workings of physiological law. It simply illustrates what nature may do in a single case under conditions exceptionally favorable.

INJUDICIOUS EXTRACTION OF PERMANENT TEETH.

A condition frequently met with after all the permanent teeth have been erupted, is one where in the upper jaw the centrals, bicuspids and molars are all harmoniously arranged, while the laterals occupy a position inside of the arch and the cuspids lie outside of it. The condition is most frequently brought about by the premature extraction of one or more members of the temporary set, as described under the last heading.

To remedy the difficulty in the easiest manner, some practitioners have at times extracted the laterals and on other occasions the cuspids. The result has been in each case an almost hopeless deformity. The cuspids brought next to the centrals oftentimes gives to the face a canine appearance, while with cuspids lacking the countenance is robbed of that prominence near the angles of the mouth so necessary to harmonious expression.

Again, the first permanent molars of one of the jaws are often neglected until caries has made serious inroads upon them, when they are extracted as offending members. The result is that the lateral pressure, so necessary to proper expansion of the process is lacking in one jaw, while in the other the normal enlargement continues. As a consequence there is disparity as to size between the two jaws, and the appearance of the individual is perhaps permanently marred.

DELAYED ERUPTION OF PERMANENT TEETH.

It sometimes happens, from causes not easily definable, that the eruption of one or more of the permanent teeth is retarded to such a degree that the rest of the set take positions in the arch and occupy all the space. When the tardy member is ready to erupt there is no place for it, and it is compelled to take a position outside or inside of the line. This is apt to occur more frequently with the cuspids than any of the other teeth, although it is occasionally met with in the case of laterals and bicuspids.

SUPERNUMERARY TEETH.

Supernumerary teeth are very frequently found occupying a position in the arch before the eruption of the permanent set, so that when the latter appear there is insufficient room for some of their number, and these are forced to assume an abnormal position. Such supernumerary teeth as appear in the line of the arch and in the anterior part of the mouth are usually small and of the conical or *peg-tooth* variety, and are most frequently found between the central incisors.

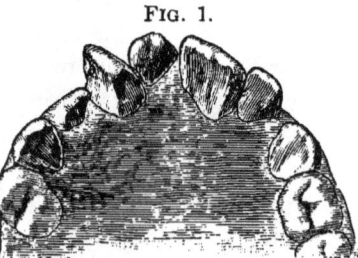

FIG. 1.

Torsion caused by Supernumerary.

Fig. 1 represents a case of this kind in the mouth of a Japanese boy, nine years of age, in which as a result of the presence of the extra tooth, the right central is turned one-fourth of a circle, while its mate is also somewhat rotated.

Fig. 2 illustrates another case very similar in character. In this instance the supernumerary tooth erupted to the side of the median line, and so only the one incisor was deflected from its normal position.

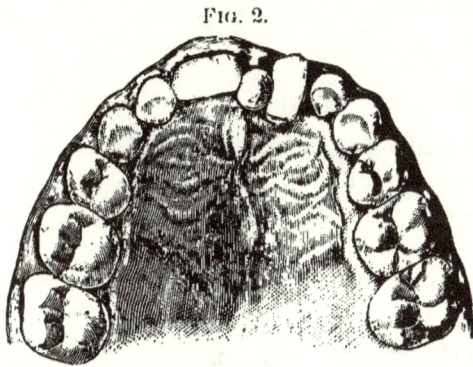

Fig. 2.

Torsion caused by Supernumerary.

Sometimes the presence of a supernumerary tooth has no other effect upon the permanent set than to occupy part of the space in the arch and separate the adjoining teeth by its own width.* Even this, however, is objectionable, for in most cases the tooth, being abnormal in form, will have to be extracted and an attempt made to close the space thus created.

ACCIDENTS.

An accidental injury to one or more of the teeth of either set, whether resulting in their loss or not, is often responsible for an irregular condition. Should a deciduous tooth become devitalized, as the result of an accident or other cause, and alveolar abscess supervene, the physiological act of absorption will be suspended, and the succeeding tooth in the course of its eruption will naturally be deflected from its course and erupt in an abnormal position.

So, also, it has happened that a deciduous incisor, through a fall, has been driven up into the process. Such a misfortune can hardly fail to cause an injury to the partially formed permanent tooth lying beneath it. Should no more

---

* A model in the Museum of the Philadelphia Dental College represents two supernumerary teeth occupying the space between the central incisors. None of the teeth are turned or misplaced, and but for the presence of these two vagrants, the dental arch would in all respects be a typical one.

serious result follow, it will probably at least divert the new tooth from its course and be productive of irregularity.

The author had one such case in his practice with an irregularly placed permanent tooth as the result.

ADENOID VEGETATIONS.

Within the past ten years the attention of oral and aural surgeons has been especially directed toward the ill effects resulting from the presence of adenoid vegetations in the naso-pharynx.

These growths are often found in children as early as the second year of life, and by partially or wholly closing the posterior nares, interfere greatly with natural breathing through the nose. They also frequently cause marked impairment of hearing by impinging upon or closing the mouth of the Eustachian tube.

It has been noticed that their presence is nearly always associated with, and by inference productive of, a pinched appearance in the superior maxillary and nasal regions of the face. This condition is believed to be attributable to lack of development of the frontal, sphenoidal and ethmoidal sinuses and the antrum of Highmore, which being normally in contact with the air, cease to develop when the circulation of the air through the nose is interfered with, resulting in altered dimensions of the face.

This lack of development in the osseous structures contiguous to the oral cavity is very likely to produce a high and contracted vault associated with a V-shaped arch, and such condition of the vault and arch has usually been found in cases where adenoid growths, through lack of discovery, have been allowed to remain through a number of years.

That these growths are directly or indirectly responsible for the malposition of teeth (as has been stated by some medical writers) other than the angular position of the superior central incisors as found in the V-shaped arch, we have no reason to believe; but inasmuch as any alteration of the normal form of the arch and vault is in itself an abnormality, manifesting itself in the facial expression, and perhaps seriously interfering with proper and needful occlusion of the teeth, it is very important that where the presence of such growths is suspected, a careful examination by means of the finger or mirror should be made, and, if found, the case should at once be referred to a specialist for surgical treatment.

HABITS.

The bad habits which young children are apt to acquire after they are weaned, such as thumb-, lip- or tongue-sucking, are important factors in bringing about an irregular alignment of the teeth in one or more portions of the arch. Acquired early, while the temporary teeth are in position and firmly set, the habit will usually make no impression upon them, but if not checked and allowed to continue up to the time of the coming of the permanent set, as is sometimes the case, these will generally be thrown out of position or so altered in their relationship as to cause a serious deformity.

This is readily accounted for when we consider that the erupting teeth, seeking their position in the arch and surrounded by newly formed and pliable alveolar tissue, are easily turned out of their course by any extraneous force exerted upon them.

The general results of the triple habit are the same, although they vary in particulars. In thumb-sucking, usually only two or three of the incisors are pressed out of place, and the ones affected are determined by the hand used and the position of the thumb in the mouth. In lip- and tongue-sucking, owing to the larger surface of the organ employed, all of the incisors will be affected.

Not only has the point of introduction of the thumb to be considered in relation to its effects, but also the angle at which it is held. When the position of the thumb in relation to the teeth forms less than a right angle, the upper teeth will be thrown out and the lower ones in; but when held in a horizontal position, the upper and lower teeth are not displaced, but simply held apart. As a result of this the first molars are kept from present contact and naturally elongate until in time they come together. The mouth is thus permanently propped apart in front, and when the second molars erupt and come into occlusion the ill-condition is confirmed. With these eight firm teeth in contact, there is no longer any hope of the ten anterior ones elongating sufficiently to meet, and we have the deformity known as "lack of anterior occlusion," which is not only a disfigurement, but a serious disadvantage to the individual in mastication and speech. This lack of anterior occlusion is not always due to the habit of thumb-sucking, for it may be brought about by physical peculiarities, as noticed fn Part III., Chapter X.

In lip-sucking the lower lip is drawn into the mouth over the lower teeth, and held there for varying periods both day and night. The result is that by the force thus exerted the lower teeth are thrown in and the upper ones out to such an extent as to give them unnatural prominence, and to cause spaces to exist between them.

Fig. 3 illustrates this condition. The child when brought to the author for consultation, was eleven years of age, and a confirmed victim to the habit of lip-sucking. Nearly all of the permanent teeth in each jaw were erupted and harmoniously related, except the eversion and introversion of the upper and lower incisor teeth respectively.

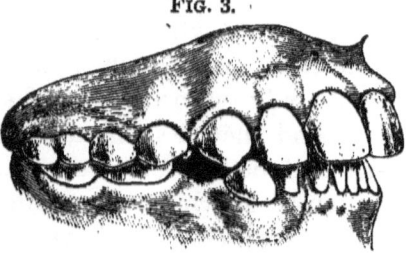

FIG. 3.

Eversion and Introversion.

The teeth were brought into proper position, and the habit, by thus being made impossible, was broken up.

The displacement and failure of occlusion of teeth in the anterior part of the mouth are, however, not the only evils associated with this habit in its three forms. In each case the jaws are held temporarily apart so that there could be no occlusion of the teeth even though they articulated normally when the jaws were closed. This leaves the side teeth free to change their position if any influence is exerted to produce that result. In the act of sucking, the cheeks are drawn in and the strong pressure thus brought to bear upon the bicuspids and (occasionally) the first molars, may cause them to be bent inward. In this mal-position they are frequently confirmed by the opportunity thus given the other molar teeth to move forward, of which they are not slow to take advantage.

IRREGULARITIES OR DEFORMITIES WITH MIXED ETIOLOGICAL CHARACTERISTICS.

There are some typical malformations of the teeth and jaws the causes of which cannot be classed under either the hereditary or the acquired form, but combine certain features of both.

Among the more prominent of these are protrusion of the upper jaw, prognathism, and the "V" and saddle arches.

SUPERIOR PROTRUSION.

In this condition the superior teeth project forward and outward to such an extent as to leave a space, more or less great, between their cutting edges and those of the lower, thus producing a marked deformity and giving to the individual a slightly imbecile expression. The lower anterior teeth, when the jaws are closed, may rest in contact with the bases of the superior ones, or they may impinge upon the gum tissue adjacent.

In some cases this deformity is but the expression of a tendency inherited from a progenitor under conditions favorable to reproduction, while in others it may be, and doubtless is, the result of mechanical forces finding manifestation in the individual alone. Even if inherited it must have been the result of such causes in the individual with whom it originated. In its acquired form, this abnormality may be caused by the slow eruption of the posterior teeth, which by failing to come in contact for a long time permit of an unusually long over-bite in the incisor region. The lower incisors thus articulating with the upper ones near their bases have a tendency to force the latter forward and outward, these movements being favored by the thin plate of alveolar process overlying the outer surfaces of their roots. As the upper teeth move outward the lower ones, from lack of restraint, elongate until their cutting edges occupy a plane considerably above that of their fellows, oftentimes fitting into and irritating the soft tissues in the roof of the mouth.

The same result is sometimes similarly brought about later in life, when through loss of several of the side or back teeth the burden of mastication is thrown upon the front ones. Lack of occlusion posteriorly and excessive pressure anteriorly will thus produce a deformity that did not exist early in life.

In some cases it may also be caused by the mal-eruption of certain of the posterior teeth, permitting them to assume a position in advance of or posterior to their normal place; such a condition would tend to restrain the lower teeth from pressing forward, and cause the upper ones to advance unnaturally.

The abnormality appears exaggerated in cases where from some cause the lower incisors incline inward, thus causing the upper ones to seem more protruded than they really are.

PROGNATHISM.

This deformity, consisting in the abnormal protrusion of the inferior teeth and jaw, is one very frequently met with.

It gives to the individual somewhat of a canine expression, and for this reason is very aptly designated by the Germans as "Hundemaul." In some cases the lower anterior teeth occlude with the superior ones, but pass outside of them, while in others the lower jaw and teeth are protruded to such an extent as to make the articulation of the lower anterior and side teeth with those of the upper jaw a physical impossibility.

Fig. 4 represents an extreme case of this character. The deformity is not only very unsightly, but interferes seriously with mastication. It is no doubt due in many cases to arrest of development of the superior arch, and is favored by any cause or causes that tend to lessen the extent of contact in occlusion. That the lower jaw possesses an inherent tendency to move forward when occlusion does not prevent is abundantly shown in cases where the individual has become edentulous and no artificial teeth are worn. Even the occlusion of artificial teeth will lessen or check this tendency.

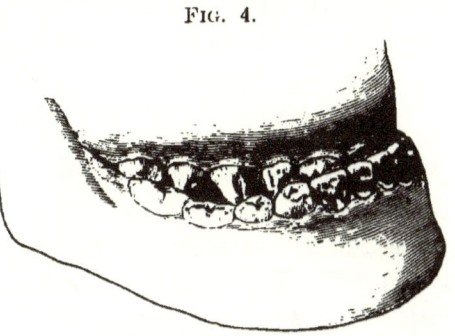

Fig. 4.

Excessive Protrusion.

In many cases it is an undoubted inheritance, while in others it may be brought about by local conditions. It is liable to occur in all cases where it is not prevented by mechanical influences.

V-ARCH.

The angular or V-shaped arch is not an uncommon one. In a typical arch of this character, the teeth instead of forming an arch, are arranged in two straight but convergent lines, which meet at an angle where the central incisors join each other. The molars, bicuspids and cuspids are usually properly related to one another, but simply thrown inward, forming straight lines instead of curves. The incisors, however, by this contraction of the space are not only thrown forward, but turned upon their axes so that their lingual surfaces present toward each other. Fig. 5* represents this form of irregularity. It is in all cases confined to the superior maxilla, the lower one being harmonious in outline. The pressing forward of the incisor teeth and their torsion often gives such prominence to the lip that the teeth remain exposed even when the jaws are closed. In addition to this unsightliness, the speech is often seriously affected by the free and uncontrollable escape of air when articulation is attempted.

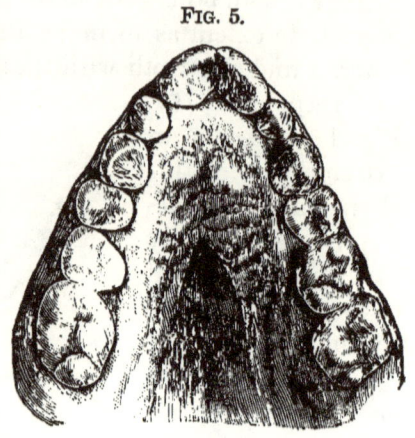

FIG. 5.

V-Arch.

The causes responsible for this condition are probably shrouded in greater obscurity than those of any other form of irregularity.

The crowding of teeth during eruption, delayed eruption or mal-occlusion, some of which are evidently responsible for many forms of irregularity, cannot be called to account for this condition, for none of them could press the teeth into such symmetrically straight lines. Mr. Charles Tomes believes that it is brought about by the pressure of the muscles of the cheeks upon the sides of the arch while sleeping with the mouth open, and that this habit is due to enlargement of the tonsils, which prevents full breathing through the nose.

---

*From a model in the collection of Dr. W. F. Fundenberg.

The pressure of the cheeks covering so large a surface would be just the kind of force likely to produce this symmetrical contraction of the arch, but we are confronted with the fact that in mouth-breathing the jaws are never held far apart, and also that the masseter and buccinator muscles, owing to their points of insertion, stand clear of the teeth, so that even when somewhat flexed, they could not possibly produce pressure upon these organs.

The condition is nearly always associated with a high and narrow vault, and it may be possible that both of these features have been brought about by imperfect development of adjacent parts, especially of the vomer, which stands in the relation of a pillar or support to the palate.

SADDLE-ARCH.

This deformity, though less common than the preceding one, and giving less external evidence of its existence, is far more likely to favor decay on account of increased surface contact. In seeking an explanation for its existence, it is well to remember that the bicuspid teeth (the ones most usually affected) are situated immediately beneath the deciduous molars, and succeed to their positions. As the first set occupies an arch in every way smaller than the permanent one, the position of the bicuspids would locate them inside of the arch described by the permanent teeth already in place. When there is no obstacle to prevent, they naturally move outward into place; but where insufficient space does not permit this, they are obliged to remain where they are, or in an effort to force their way into line, assume a crowded and irregular position.

The fact that when bicuspids are out of line they are nearly always found to be inside of the arch seems to favor the supposition that the irregularity has been brought about in the manner suggested. Early eruption of the cuspids and tardy eruption of the bicuspids would also favor the condition.

The assumption that bicuspids once in line may be forced out of it by pressure exerted in the eruption of the second and third molars has little to support it. Were this possible or probable the deformity would be more frequently met with. Fig. 6 is a fair representation of this deformity. Both sides of the arch are not usually affected to the same extent, and in some cases the two bicuspids on one side occupy a position directly across the arch, each one being partly turned upon its axis. The condition is rarely met with in the lower jaw, and is one, according to the author's observation, never inherited, but always acquired.

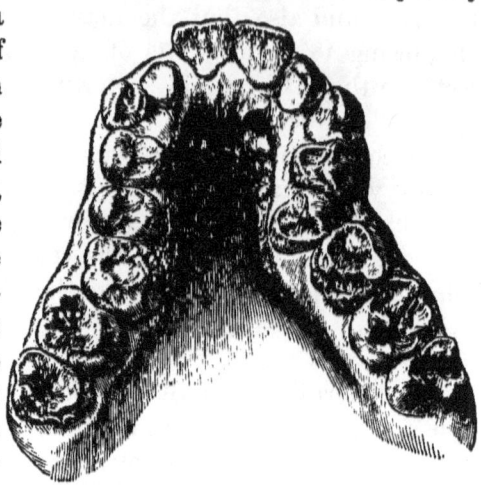

FIG. 6.—Saddle-Arch (*Coleman*).

CHAPTER III.

EVILS ASSOCIATED WITH IRREGULARITY.

In order to properly appreciate the importance of correction of irregularity of the teeth, it will be well to consider in brief detail some of the more prominent evils associated with the condition.

APPEARANCE MARRED.

While this result is usually not the most important of those connected with irregularity, it is the one which most generally induces the patient to apply for remedial treatment. The other evils may not be recognized, or may be considered of minor importance by the parent, but the ill-appearance of the child both attracts the attention and enlists the sympathy to such an extent as to create a desire for its improvement.

The external deformity caused by an irregularity will be greater or less according to its extent and location. If it be slight in character and located back of the cuspid teeth, it will usually give no external evidence of its existence, but if located in the anterior part of the mouth, it will, even if slight, be very noticeable and in consequence constitute a source of annoyance to the individual throughout life.

The class of irregularities most noticeable under all conditions is that where the form of the arch is altered, thus changing in a marked degree the entire facial expression. Such deformity cannot be masked. It must either be mechanically reduced or stoically endured.

SPEECH AFFECTED.

This result like the preceding one will be slight or aggravated according to circumstances, but when at all considerable it proclaims itself to the world with every attempt at speech in so unpleasant a manner as to be a painful annoyance to both speaker and listener.

It may be due to the restriction of the movements of the tongue as in a narrow or contracted arch, to alteration of the form of the roof or vault of the mouth where the sides of the latter have assumed a deep pitch resulting in the formation of a sharp angle along the median line of the palate, or, it may be, and most usually is, due to the uncontrollable escape of air between the teeth in the anterior part of the mouth by virtue of the non-approximation of those teeth and the change of form in that part of the alveolar ridge which aids the tongue in the production of perfect sounds.

MASTICATION IMPAIRED.

In most cases of irregularity, either simple or complicated, there is a corresponding degree of either mal-occlusion or lack of occlusion. In simple cases, or where but few teeth are thrown out of occlusion, it may not occasion any inconvenience to the individual, but where the irregularity is at all extensive so many teeth are usually lacking in occlusion as to seriously impair the power of mastication.

When this latter condition prevails it is most likely to result, sooner or later, in injury to other organs, for where mastication is imperfectly performed greater demand is made upon the stomach to prepare the food for digestion and assimilation. The stomach soon feels the effect of this over-taxation and becomes weakened in tone, which may finally result in incapacitating it for the performance of its normal functions.

Teeth that do not occlude are of no use to the individual for purposes of mastication, and those that occlude but slightly or imperfectly possess very slight value.

As one of the principal functions of the teeth is mastication, and as all the teeth are needed to perform this work satisfactorily, it naturally follows that any interference with this function, through irregular position or otherwise, must be detrimental to the individual and may result in partial or complete loss of health.

CARIES INDUCED.

The human teeth are arranged in the jaws in such manner as to best subserve the wants of the individual, and their form and location are also such as to conduce to the greatest immunity from caries and their consequent longest endurance.

Their rounded approximal surfaces and the constriction of their necks reduces the points of contact with their fellows to the minimum. As their liability to approximal decay is in proportion to the amount of surface in contact, it will be seen that those normally placed are likely to be freest from the ravages of caries.

When, therefore, the teeth occupy irregular positions, especially where they are crowded, more of the surface of each tooth is in contact, and the liability to decay is correspondingly increased. This is true of irregularly placed teeth in any part of the arch, but the liability is greatly increased where crowding or overlapping exists among the incisor teeth, for owing to their flattened form it is possible for more of their surface to be in contact with their fellows than would be possible with any of the other teeth. In such cases, with the condition uncorrected, teeth decay and re-decay, in spite of the most faithful efforts of the dentist, until they are finally lost.

CHAPTER IV.

ADVISABILITY OF CORRECTION.

With our present knowledge in regard to the teeth and their surrounding tissues, and the advancement made of recent years in the multiplication and perfection of mechanical appliances, scarcely any deformity of the mouth and teeth is beyond mechanical remedy. With possibility assured, however, it is most important that we should consider carefully the question of advisability, for what is possible may not always be advisable. There are several considerations that enter into this question of advisability.

AGE.

The age of the patient has much to do with the advisability of any proposed operation for correction. Early in life, when the alveolar tissues have not yet reached the hardness and density of structure which they will attain at a later period, they are more easily operated upon. They are elastic and readily yield to pressure, and at the same time under the influence of this pressure they are more quickly resorbed or bent and thus give way to the tooth that is being moved. This feature of early youth is an important and valuable one in that it renders an operation for correction more easy of accomplishment, but while the soft and easily yielding process favors the operation, it is at the same time a tissue poorly fitted to resist the influences which often operate to again displace the tooth. For this reason, a tooth moved at an early age may be liable to subsequent displacement when the pressure caused by the eruption of the succeeding teeth is brought to bear upon it.

After maturity, we have the conditions exactly reversed. The denser and more perfectly calcified process yields less readily to pressure and absorption, but when the tooth has once been moved into proper position it is more easily and firmly held there by the surrounding tissues.

In view of these facts it will readily be seen that in many cases, especially where the proposed operation is simple in character, and where the result obtained is not likely to be nullified by subsequent events, interference early in life is advisable, but where the operation is to be extensive in character, and especially where we have reason to doubt our ability to retain the results secured, prudence would suggest non-interference until all of the fourteen teeth of the involved jaw have erupted.

HEALTH.

The health and strength of the patient at the time of any proposed operation for irregularity is so important a consideration that it dare not be disregarded. The time that is generally considered most favorably for correction (between the ages of thirteen and eighteen years) is also a period when important changes are going on in the entire economy. The individual is passing from the stage of childhood into that of manhood or womanhood, and in this change, especially in the case of the female, the life-forces are taxed to the utmost. At this time also the mental faculties are being severely strained by study, in consequence of which, if the physical culture of the individual be neglected, as it too often is, the nervous system becomes unduly exalted.

To meet and partially compensate for these drains upon the system, it is most important that full nutrition be sustained. To do this with teeth that are sore or tender to the touch from being moved is impossible, and hence the system will be still further weakened by lack of nourishment if any severe operation be undertaken.

At this period of life, therefore, unless the patient possesses vital powers of a high order, it might be unwise to further tax his or her system by any extensive operation for correction that would involve the infliction of much pain, discomfort or annoyance. Should the vitality of the patient be below the average, no difficult or protracted operation for correction should be undertaken, for it might result in permanent impairment of the health.

It is much better to postpone the operation until a time when the vital powers can stand the strain or if necessary abandon it altogether, for the loss of health can never be compensated for by any benefit conferred upon the dental organs.

SEX.

The sex of the individual must also be considered in connection with this subject. The consideration of sex may be disregarded so far as the desirability of an operation is concerned—for if the results of neglected irregularity are harmful in respect to one sex, they are certainly equally so in regard to the other—but as regards the necessity for interference the question of sex is an important one. Correct facial expression and harmony of feature are far more important to the female than to the male; for, being endowed by nature with greater beauty of form and feature than man, its absence in any part is more noticeable than it would be in the sterner sex. Besides this, after youth is passed, man has in the hairy covering of the lip a means of concealing most deformities of the dental arch, while woman is entirely without this advantage. For these reasons the necessity for the correction of any irregularity of the teeth seems more imperative in woman than in man.

POWER OF APPRECIATION.

The intelligence of the patient and his ability to properly appreciate any benefit conferred, are important considerations in enabling us to determine whether or not to undertake any considerable operation for the correction of irregularity.

Correction of irregularity is at best a most difficult undertaking, and frequently lacking in suitable pecuniary reward, so that the lover of the art must nearly always depend upon appreciation for part of his compensation. If this be wanting, the operation is robbed of nearly or quite all of its attractiveness, and the stimulus to success is absent.

There are those whose want of intelligence or lack of culture would lead them to regard with much indifference any irregularity of their teeth, and who if benefited by our efforts for correction would fail to appreciate it. For such it would be manifestly unwise to urge or encourage any difficult or extensive operation for correction even though they might be able to compensate us pecuniarily for our labor, for they would be likely either to give up the operation when partially completed or fail to wear any appliance for retention, and thus permit failure to follow success.

FAMILY TYPE.

When any great deformity of the teeth and jaws, such as anterior protrusion of either jaw or a V-shaped arch is shown to be hereditary, it is well to take into consideration the hereditary feature of the case before beginning any operation for correction. Where the irregularity is known to have been acquired in the parent of the child and thus to have been transmitted but once, the difficulties in the case are not so marked because the type has scarcely been confirmed; but where it has been transmitted through two or more generations the impress is strong and difficult to overcome.

In the latter case the correction of the deformity will not be more difficult than usual, but after correction the tendency of perverted nature to cause a return to the family type will be so strong as to almost baffle us in our attempts to preserve the advantage we have gained. Under such circumstances the retaining appliance will have to be worn a very long time, and a constant watch kept over the case until we are sure that the result will be permanent.

IMPROVEMENT OF OCCLUSION.

Faulty occlusion is always necessarily associated with irregularity and is one of its most objectionable features. While mastication may be performed to the satisfaction of the individual where an irregularity exists, it can neither approach the ideal of nature nor properly subserve its own ends unless the teeth articulate in a normal manner. It would be difficult to find a set of teeth in which the articulation is all that could be desired, but the nearest approach to it is what we should strive after. Therefore, in considering the advisability of correction in any given case, we should carefully study the existing conditions and endeavor to ascertain in advance whether, when we have improved the arrangement of the teeth, we have also improved the articulation.

Dr. Davenport says: *"In the treatment of our patients, it is hoped that if we cannot all see our way clearly upon this matter, we may at least see far enough not to make the articulations worse by our operations than they were when brought to us."

"Much harm is done by the use of regulating appliances which change the articulation without improving it, and it is almost a universal fact that unless an improvement can be made in the articulation, there will be no permanent improvement in the irregularity."

---

* International Dental Journal, Jan., '92.

CHAPTER V.

AGE AT WHICH CORRECTION MAY BE BEGUN.

The correction of irregularities, under favoring conditions, may be begun and carried forward successfully through a wide range of years.

It may be undertaken as early as the eighth or ninth year, and again may yield successful results as late as the thirty-fifth year or later. The operation is one largely dependent upon the absorption and re-formation of alveolar tissue, and as new bone will form at almost any period of life, as evidenced by the reunion of a fracture, so the correction of an irregularity is possible at quite a late period.

The correction of irregularity, however, would usually prove so slow and tedious an operation after the maximum of density had been attained in the process, and the necessity for it be so much lessened by advancing age, that the advisability of undertaking it would be questionable.

WHEN EARLY INTERFERENCE IS JUSTIFIABLE AND ADVISABLE.

Any of the permanent teeth may erupt outside or inside of the arch. If allowed to remain in such position for any length of time, the space intended for their accommodation will soon be partly occupied by the adjoining teeth, and the subsequent correction of the irregularity rendered more difficult. So also a central or lateral incisor often erupts in such a manner that its cutting edge, instead of being in line with the curve of the arch, forms an angle with it.

This torsion may be associated with an overlapping of the adjacent tooth as shown in Fig. 7, or there may be a space between the two as shown in Fig. 8.

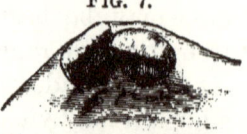

Torsion and Overlapping.

In either case the twisted tooth occupies a less space at the line of the cutting edge than it should. By allowing this condition to remain, when the pressure of the later erupting teeth begins to be felt, these teeth will be pressed still closer together and the irregularity be confirmed. Subsequently, when the correction of the condition is attempted, there

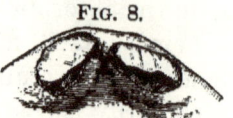

Torsion with Space.

will not be sufficient room to accommodate the tooth in its wider aspect and the adjoining teeth will have to be pressed apart or the arch expanded to obtain the necessary room; whereas, if the tooth had been turned in its socket before the eruption of the other teeth the operation would have been a very simple one.

Again, when an incisor erupts so as to occupy a position inside of the arch in the upper jaw, or outside of it in the

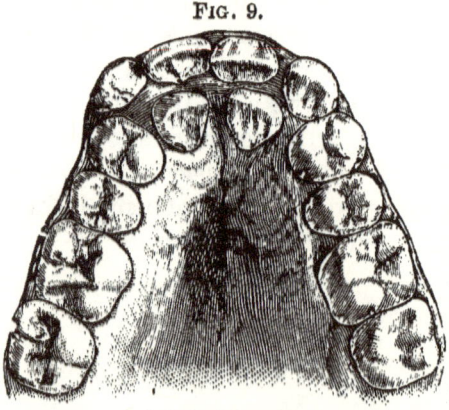

Inlocked Laterals.

lower, and the tooth be held in such position by the antagonizing teeth, immediate interference and correction is demanded in order to prevent the complications that would result from the partial or complete closure of the space intended for the accommodation of the malposed tooth. Fig. 9 shows a case of this character with both laterals inlocked. The superior central incisors sometimes erupt in such a manner that their cutting edges form an angle at the median line. To neglect the condition or to postpone its correction would not only result in its confirmation and probable aggravation, but might also open the way for a complete change in the shape of the arch. Fig. 10 illustrates this condition.

It is entirely probable that certain arches of a modified V-shape have been formed in this way. In cases such as those just mentioned, early interference is the wiser plan, but it is equally important that after they have been placed properly in line they should be firmly held, not only until new bony tissue has been formed around them, but until the lateral pressure of the neighboring teeth coming into place has spent itself.

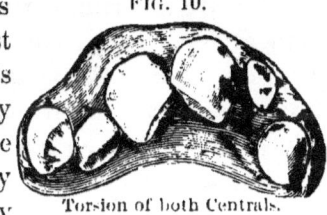

Fig. 10.
Torsion of both Centrals.

How this may be readily and successfully done is shown in the consideration of practical cases in Part III.

In the lower jaw the conditions are somewhat different. The incisors, upon eruption, generally assume a somewhat crowded and irregular position, which is partly or entirely corrected by nature in the enlargement of the arch and the influence of the lip and tongue in bringing them into a more harmonious outline.

Interference with them when they are within the arch is not usually called for until a later period.

While there are many practitioners who have long held the view that early interference is inadvisable in the great majority of cases, the large experience of others who have made a specialty of this branch of practice has led them to declare in favor of early correction.*

---

* "As soon after eruption as it becomes certain that an irregular denture is inevitable, there is no longer justification for delay, and after that period every year increases the difficulties, both mechanical and pathological, and prejudices the stability of the dental apparatus. . . . The author has not hesitated to undertake treatment of very extensive irregularities, even while teeth were emerging from the gums."—Kingsley—Oral Deformities, pp. 61, 62.

Certainly where the case is brought to our attention in its incipiency we can frequently by judicious management and the application of slight corrective measures either counteract or greatly modify any tendency toward irregularity.

A feature favoring early correction is that the roots of the teeth are not fully calcified until a long time after their eruption. Until fully calcified the apical foramina are large and more than accommodate the nutrient vessels entering the tooth, so that there is less danger of devitalizing the pulp through strangulation in the movement of a tooth at this period than later.

WHEN CORRECTION SHOULD BE DELAYED UNTIL DENTITION IS COMPLETE.

In some cases an extensive operation for the correction of irregularity involving a number of teeth, should not be undertaken until all of the permanent teeth (excepting the

---

"It is the opinion of the writer that each tooth should be encouraged to take a correct position in the circle of the arch while erupting (or as soon thereafter as practicable), in order to promote the proper development of the jaw, for the teeth next to be erupted are thus more likely to do so in proper position and order."—Jackson—Trans. Amer. Dent. Assoc., 1890, p. 201.

"I believe the best time to begin the treatment is as soon as the appearances of irregularity are manifest, then, with delicate and simple appliances, gradually assist the tooth to take its natural position. . . . A few days with a proper appliance will often accomplish what may require many months if left until the whole dental apparatus is involved."—Angle—Pamphlet, p. 50.

"Early interference is often necessary, as where the superior incisors erupt slightly posterior to their natural position and occlude with the cutting edges of the lower ones. Should the superior incisors not be moved forward as soon as this tendency is noticed permanent prognathism might result."—Goddard—MSS.

"As soon as a tooth (or a number of teeth) erupting show such misplacement that natural conditions will not make it self-correcting, mechanical means should at once be resorted to—no matter what the age. The only condition that would not justify such interference is ill-health."—Matteson—MSS.

third molars) are fully erupted. When but a few teeth are malposed with no prospect of their being able to take their places in the arch unaided, and every prospect of their being confirmed in their malposition, the necessity for immediate interference is plainly evident; but where a large number of teeth are malposed it is not so easy to prognosticate what effect their correction may have when considered in relation to the teeth still to be erupted. The result is naturally involved in some doubt. Even if the necessity for correction appears evident to us and we should accomplish it, the final result may not be all that we had hoped for.

Under such circumstances it is wise to delay interference until the permanent teeth are in place and the arch fully expanded, when by a careful examination of all the conditions we can easily foresee the result of any proposed operation and decide intelligently not only what needs to be done, but also the best way of accomplishing the desired result. Oftentimes this later examination will show that the irregularity has much improved and the necessity for interference is consequently lessened.

The line of distinction between the advisability of early and late interference is not always plainly marked, but where there is no very evident reason for delay prompt interference is the safer and better plan.

CHAPTER VI.

MOVEMENTS TO BE PRODUCED AND PRINCIPLES GOVERNING THE APPLICATION OF FORCE.

In causing malposed teeth to assume their proper positions in the arch certain movements are necessary, and to properly accomplish them force must be brought to bear in a manner best calculated to produce the desired result. The usual movements that teeth undergo in being forced into position, are outward, inward, forward, backward and rotary. Sometimes but one movement is necessary in the case of a single tooth, but more frequently several are required before proper alignment is secured.

The application and regulation of force in producing movements of the teeth are governed largely by the general principles of applied mechanics.

The greatest good can be obtained from any force only when it is exerted in a direct line with the movement desired.

To this end, in the selection or application of any devices for the moving of teeth, preference should be given, *caeteris paribus*, to those that are most direct in their action.

The application of direct force, however, is not always possible owing to the position the power-producing instrument would have to occupy in the mouth, and the consequent interference (as in the lower jaw) it would cause in limiting the movements of the surrounding or adjacent organs. For this reason we very frequently have to consent to the use of some form of appliance that will yield power in a line that is not direct, but still effective.

The force used must be sufficient, but not excessive, and not too abruptly applied.

If the force be insufficient to accomplish the desired object the result will not only be a failure, but it will also involve a serious waste of the time of both patient and operator; whereas, if it be more than sufficient it might cause a fracture of one of the alveolar plates, a rupture of a blood-vessel at the apical foramen, or a constriction of the entire pulp at the same point, resulting in its devitalization.

So, also, in the widening of the arch, if the force be too great or too suddenly applied it is liable to result in the separation of the superior maxillary bones at the palatal suture. While this result need not necessarily be harmful, if occurring to a limited degree only, it is one that usually occurs without our wish, and in all cases requires suspension of further operations until the parts have again united. The greatest prudence and care are necessary in the application of force to the teeth.

The points of resistance and delivery of the force must be fixed points.

The point of resistance, or in other words the point selected to resist the strain of an appliance while it is being exerted to cause movement at some other point, must necessarily be fixed and immovable, for if it be not so, fully one half or more of the force expended will be lost. Besides if the anchor tooth or teeth should yield at all to the pressure they would be pressed out of place and thus one irregularity would be created in our attempt to correct another.

No factor in orthodontia is more important than this.

So, also, the point of delivery must be a fixed point. By a fixed point in this sense, is meant one that will receive the force in such way that none of it will be lost. As the tooth is moved, the point on its surface where the force is delivered will necessarily move with it, but it should be so arranged that in this movement the point of delivery be not changed. A change at this point will be as disastrous, and frequently more so, than at the point of resistance, for if the appliance should slip or change its position, the force would be exerted in a line different from that intended and harm result.

Great difficulty was formerly experienced in making attachments for appliances so that they might be immovably held where placed, but since the introduction of the metal band or ferrule by Dr. Magill, difficulty of this character has been overcome.

The resistance at the point from which we exert pressure must be greater than the resistance to be overcome by the pressure.

The truth and importance of this statement would seem to be self-evident.

Our points of resistance usually consist of one or more teeth situated at some distance from the one intended to be moved. Occasionally, a single tooth, if it be multi-rooted or one with a long root firmly implanted, will be sufficient for our anchorage, provided the tooth to be moved be single-rooted and of not too great resisting power; but a tooth with a single root will seldom be sufficient for anchorage in moving any other tooth. A single molar, firmly implanted, may sometimes be sufficient to offer resistance in the moving of a bicuspid or incisor, but it is always better, if possible, to have the resistance divided among several teeth.

A cuspid should never be depended upon to resist alone the force needed to move another cuspid, for it is as likely that the one will be moved out of as the other into place. The force of resistance should always be as much distributed as possible, for the sake of safety.

It should always be seen to, in advance, that there is sufficient space to accommodate the tooth in the new position it is to occupy.

The importance of this precautionary measure will be readily seen. Unless there be room to accommodate a tooth we will either fail in our efforts to move it or succeed only by the expenditure of an amount of force out of all proportion to the requirements of the case. Instead of moving one tooth we may under such circumstances have to move several at the same time, a difficult and oftentimes unnecessary undertaking.

If sufficient room does not exist naturally, we can increase it by separating the adjoining teeth. If the space already existing be too great to admit of the use of rubber wedges, the object can be accomplished by the use of compressed wood, or other suitable substance.

In certain favorable cases an inlocked tooth may be brought into place and room provided for it at the same time by the use of the combination band and screw (or spring) appliance shown in Part 3, Chapter III.

In many cases where a tooth is locked out of place the jaw needs or will bear expansion as well. In such cases, of course, we expand the arch first, and this will afford us room to bring the tooth into position.

An exception to this rule is sometimes found in the case of a lower incisor placed slightly within the arch and held there by the adjoining teeth. As these teeth are usually easily moved it will not be necessary to provide room in advance, for, if our point of resistance be sufficient, we can, by the use of a jack-screw, readily force the tooth into line, notwithstanding the overlapping of adjoining teeth. An illustration of this method is shown in Part 2, Chapter 1.

Where expansion of the superior arch seems to be indicated satisfy yourself that it is the best course to pursue before undertaking it.

The advisability of expanding the superior arch must be determined by existing conditions. If the lower arch be larger and more prominent than the upper one, prevention of further enlargement of the lower one by extraction of one of the molars on each side may be better practice than enlargement of the superior arch. A careful consideration of the facial expression as shown in the harmony of its various parts will help us to decide the matter.

Again, the size of the superior maxilla should also be a determining factor. To spread the superior teeth upon a bony arch of moderate or inferior size in order to make them meet or overlap the prominent teeth in the lower jaw would cause them to slant outward to such an extent as to produce in itself a very decided deformity.

The relation of the maxillary as well as the dental arches should be carefully studied before any plan of treatment is decided upon.

Pressure may be either continuous or intermittent.

The question of the use of either continuous or intermittent pressure in the regulation of teeth did not arise until Dr. Farrar declared, a number of years ago, that, according to physiological law, direct and intermittent pressure was the only kind suitable to be applied in the moving of teeth.

The only way in which direct and intermittent force can be applied is by the use of the screw in one of its various forms. Continuous pressure is that which we obtain from the elasticity of the metals, from rubber either partially or fully vulcanized, and from the expansion of wood, sea-tangle or other like substances. The action of these substances cannot well be interrupted to provide a period of rest, but they continue their action until the force they are designed to exert has been spent.

The screw is, in most cases, one of the best methods by which to exert pressure, but it cannot be applied to advantage in all cases. To limit ourselves therefore to its use, would be to deny ourselves the advantage to be gained by the employment of the various substances previously enumerated.

So far as the author is aware no one has advocated the exclusive employment of continuous pressure, but those who believe in and make use of it, also avail themselves of intermittent pressure in the form of the screw, not because of its interruptability, but because of its directness and power.

Experience has shown that by continuous pressure equally good results have been produced as by interrupted pressure and with as little harm. Those who, like the author, have used both kinds according to the seeming requirements of the case in hand, have been unable to notice any advantage in the one over the other as viewed from a physiological standpoint.

Dr. Atkinson once expressed his belief that continuous pressure in regulating most fully stimulates the action of the osteoclasts in the absorption of alveolar tissue.

Pressure should be exerted as nearly as possible in a line at right angles to the long axis of the tooth.

By the application of power in this way the best results are accomplished. If the power be applied at a slight angle from above, no harm will result, as it will only serve to keep the tooth in its socket while it is being moved, but if applied at an angle from below, the tendency will be to lift the tooth from its socket and serious complications may ensue.

This last result is most liable to follow the use of a jack-screw applied at an improper angle, when by its direct and excessive power the tooth may be elevated and partially dislodged.

Pressure once applied should never be wholly relinquished until the operation is completed.

When from any cause it is desired to suspend the movement in progress, it may be done by temporarily ceasing to exert pressure, but the appliance should be kept in position to retain the advantage gained. If removed, the teeth would fall back toward their old positions and great soreness and inconvenience result.

CHAPTER VII.

EXTRACTION AS RELATED TO ORTHODONTIA.

Probably no feature in the practice of Orthodontia is more important, or has associated with it greater possibilities for good or evil to the patient than that of extraction.

As related to the prevention or correction of irregularity, extraction on the one hand may be of the greatest possible benefit or on the other it may result in irreparable injury.

Judicious extraction, if undertaken in time, will often forestall or prevent an irregular condition of the teeth, and in other cases it will assist greatly in simplifying the operation of correction. Occasionally, it is all that is called for on our part, nature performing the rest of the operation unaided.

Injudicious or ill-advised extraction, however, may complicate and render most difficult the correction of cases which in themselves were not difficult, or it may even be the immediate cause of a deformity which would not otherwise have existed.

The paramount importance, therefore, of knowing when to extract, and when not, will be readily recognized.

To properly convey to the student a fair understanding of these circumstances, in as concise and comprehensive a manner as possible, it has been thought best to formulate the following rules:

Avoid, if possible, extracting any of the six anterior teeth in the superior arch.

We would urge this, because it is nearly always unnecessary to extract them, and because their absence, owing to their prominent position, would be more noticeable than that of other teeth in the mouth. If the anterior teeth be sound and only irregular in position, the extraction of a bicuspid from one or both sides will usually give us sufficient room for spreading the anterior teeth and moving them into their proper positions.

It has happened, however, to the author and others, to meet with cases where the superior laterals were locked inside the arch by the close approximation of centrals and cuspids, and where the laterals were withal so badly injured by decay and disease as to render their usefulness doubtful if brought into line. In such few cases it was deemed best to extract the laterals, especially as their absence would not be more noticeable afterward than before, and because there was good occlusion between the rest of the teeth in the mouth.

The author had two unusual cases present to him in one year for the reduction of protrusion of the anterior superior teeth. In each case there was a broken or badly diseased right central incisor that was beyond hope of preservation. In these cases it did not happen particularly amiss, for the extraction of the roots afforded room for drawing in the remaining five teeth, thus easily reducing the deformity and at the same time closing the space. The appearance of the patient in each instance was greatly improved, and the absence of even so large a tooth as the central was scarcely noticeable.

It must be borne in mind that in the cases just mentioned advantage was simply taken of an existing condition to simplify an operation. Had the teeth been good, the proper plan to pursue would have been to extract a bicuspid on each side and retract the anterior teeth.

In another case, a girl eleven years of age had lost a right superior central incisor through a fall from a swing. Two days after the accident, and when the tooth had been mislaid or thrown away, she was brought for treatment. Only two methods of remedying the difficulty suggested themselves. One was the wearing of an artificial tooth; the other, drawing the teeth together to close the space. The latter plan was decided upon, and successfully carried into effect; but, unfortunately, as there had been no protrusion before and there was contraction afterward, the superior teeth no longer overlapped the lower ones, but met them edge to edge, thus giving the upper jaw a flattened appearance, which was in itself a deformity. The patient was saved the annoyance of wearing a plate, but her facial expression was injured in consequence.

Such cases as those just alluded to are exceedingly rare, and are only mentioned as extraordinary exceptions to a very good rule. Aside from the centrals, there is probably less excuse for the extraction of the cuspids, than any of the anterior teeth, and yet it is, unfortunately, too often resorted to.

If for any cause the cuspids erupt abnormally, and there is no room for them in the arch, if it be not advisable to expand the arch, one of the teeth on each side should be extracted to make room for them.

The question of which tooth to extract (where extraction is deemed best) in order to allow an outstanding cuspid to be brought into line will be largely governed by the position that this tooth occupies. If it be situated in a line between the lateral and first bicuspids with its root sloping backward, the bicuspid would be the proper tooth to extract. If, however, the slant of the root be forward the lateral incisor should be removed, because in drawing the crown of the cuspid forward to occupy the place of the lateral, it would be made to occupy a more nearly vertical and normal position. To move the crown of a cuspid forward when it already inclines forward, or backward when it points that way, would cause it to present such an oblique appearance as to be very unsightly.

In the lower jaw one of the incisors may sometimes be extracted to gain space.

Slight irregularity or crowding of the inferior incisors is of such common occurrence as to have almost become the rule instead of the exception. Their partial concealment, together with the usual freedom of the condition from ill results, causes any interference to seem meddlesome rather than otherwise, if the irregularity be trifling. In cases, however, where the crowding is excessive and calls for correction, it is usually the easier and better plan to extract one of the implicated teeth and bring the others together into line. The four teeth are so nearly alike in size and character, that the loss is not usually noticed when one has been removed. It is sometimes perplexing to decide which of the four to extract, but the one most out of line, and in consequence the one that will create the least space by its removal, should usually be selected.

In respect to the loss of the inferior cuspid, the same remarks apply as to its fellow in the opposite jaw.

Back of the anterior teeth, if all are equally good and one must be removed, select the one nearest and posterior to the one out of position.

As so large a proportion of the irregularities we are called upon to correct pertain to the anterior teeth, and as it is so important to retain these, extraction for room, when necessary, generally falls upon one of the teeth posterior to the cuspids. Which of these it is best to extract, to make room for a malposed cuspid or incisor, has been a subject of controversy among practitioners for many years.

Some have claimed that as the statistical tables show the first molar to be by far the least durable of all the permanant teeth, it should generally be selected as the one to be sacrificed. Others, on the contrary, have contended that as the first and second bicuspids are both frail teeth, and are often lost early in life, and as from its greater size the first molar is so much more valuable in mastication, it should be preserved and one of the bicuspids removed.

There is truth in both of these arguments, but we feel

satisfied that under the conditions named, (all equally good at the time,) wisdom will dictate the removal of the one nearest the point of difficulty, for in so doing we greatly simplify the operation for correction and effect a saving all around. Simplicity in surgical as well as mechanical matters is a great desideratum. Indeed, it not infrequently happens that where a cuspid is out of line the first bicuspid has usurped its place in the arch, so that if we were to extract the first molar, both first and second bicuspids would have to be moved out of their position of good occlusion into a space further back, a feat very difficult and oftentimes well-nigh impossible of accomplishment. By the simple extraction of the first bicuspid in such cases, the cuspid will usually fall into its place without any assistance.

In certain cases where there is nearly but not quite sufficient space in the arch to accommodate an outstanding cuspid, and where the occlusion would not contra-indicate such action, it is better to extract the second bicuspid instead of the first, because the surplus space thus created will then be back of the first bicuspid, and consequently less noticeable than it would be in front.

If a tooth other than the one nearest to that in malposition be defective, and not too far distant from point of irregularity, extract it instead.

The second molar, decayed or sound, is usually too far distant to be available by its extraction in furnishing room for the movement of anterior teeth. If the bicuspids be sound and the occlusion does not interfere with their backward movement, the first molar, if very defective, may be extracted in preference to a sound tooth in advance of it.

So, too, if the second bicuspid be carious or defective and the first one healthy, the former should for the same reason be extracted.

If a tooth must be lost, either to allow a more important one to fall into line or to create space, it should be done without delay to accomplish the best results.

When a cuspid erupts without room in the arch for its accommodation, and the circumstances of the case point to the extraction of the first bicuspid to make place for it, the sooner the extraction takes place the better. If the operation be delayed, the cuspid in its endeavor to force its way into place will often press so hard upon the lateral as to force it inward, and if possible under the central, thus creating an additional irregularity. Such results have often been noticed. Prompt extraction after it had become necessary would have changed the condition.

In similar manner, when it becomes advisable to extract one or more of the first molars to prevent the further expansion of the jaw or to abort a threatened irregularity in the anterior part of the arch, it is best not to delay their extraction too long. They should not be extracted before the second bicuspids are in place, but if they must be lost, they should be removed after the eruption of the latter teeth and before the second molars appear, somewhere about the eleventh or twelfth year. If longer delayed the harm we wished to prevent (expansion of the jaw) will have been accomplished and their later extraction will not avail. If extracted about the time the second molars are erupting, the latter will glide naturally into the space formerly occupied by the extracted teeth; this they are not so apt to do later.

If a tooth must be removed on one side to obtain space it does not necessarily follow that its mate on the opposite side should also be extracted.

If there be the same reason for extracting both, as where the existing evil pertains as much to one side as to the other, let both be extracted; but where the trouble sought to be remedied is confined to one side, the extraction of a tooth on that side ought not to be supplemented by a useless extraction on the other. Those who favor symmetrical or double extraction claim that it prevents the disturbance of the median line, but it has been our experience that the extraction of a tooth back of the cuspid will not often affect the central line through the moving of the teeth toward the space, and even a slight disturbance of that line is far less objectionable than the sacrifice of a valuable tooth. A correspondent mentions a case in which, after a long struggle to save the badly decayed superior first molars of a Miss, 14 years of age, he determined to extract them. After three months the girl returned, and it was noticed that the superior centrals had separated one-eighth of an inch. He adds, "since that time I have refused to extract symmetrically in-growing subjects."

Where there is disparity in size between the two jaws, and two teeth need to be extracted from the more prominent one, it would be a serious mistake to extract the corresponding teeth in the other and smaller jaw.

It would seem almost impossible to make such a mistake, and yet that it has been made time and again, the mouths we are called upon to examine often bear sad evidence. It occurs through lack of knowledge, want of judgment, or erroneous teaching.

When those of long practice advise, without qualification, that at eleven years of age the four first molars should be extracted, it is scarcely to be wondered at that some young practitioners should lose confidence in their own better judgment and be led astray. Harm of this nature, when once done, can never be undone, and the patient is injured beyond remedy.

Needless extraction should be carefully guarded against.

It is our object to save and improve, not to destroy. Extraction should only be resorted to when it appears, after careful consideration, to be the best or only way of accomplishing the object in view. Ill-advised extraction of the molars or bicuspids has often been the cause of a very serious and irremediable form of deformity, namely:—the separation of the anterior teeth, leaving unsightly spaces between them, thus depriving them of their natural support and leading to their earlier loss.

When teeth, especially the first molars, are extracted at a later period than they should be, leaving a space that the second molars cannot occupy, the teeth anterior to the space will fall back unless prevented by the occlusion. If this falling back pertains only to the bicuspids, no harm will usually result, but if it extends to the anterior teeth, as it may, and often does, the result will be disastrous. In this connection we cannot help again emphasizing the necessity for the removal of first molars (if they are to be removed) before the second molars have assumed their places in the arch.

CHAPTER VIII.

PHYSIOLOGY OF TOOTH-MOVEMENT AND CHARACTER OF TISSUES INVOLVED.

In changing the position of teeth in the act of regulating, the surrounding tissues, both hard and soft, are largely involved.

In order, therefore, to properly comprehend the philosophy of tooth movement, it is necessary to understand the structural character of these tissues and the physiological changes that take place in them while a tooth is being moved.

THE ALVEOLAR PROCESS.

This process, as its name implies, is not a separate and distinct bone, but an outgrowth from another. It is a provisional structure designed to support the teeth in position and afford lodgment for the nutrient vessels leading to them. It is formed upon the body of the bones of the jaw as the teeth are developed, growing with them until they are fully formed, and then remaining while they remain.

When the teeth are lost, there being no longer any special use for it, most of this process is absorbed and carried away. In early infancy little alveolar structure exists, but it is formed co-ordinately with the growth of the deciduous teeth and remains during the period of their retention. Should they be lost before their successors are ready to appear, the process will be entirely removed by absorption, and a new one formed for the accommodation of the permanent teeth. Where, however, the deciduous teeth are gradually shed to make way for their successors, the process is not entirely absorbed, the basal and unabsorbed portion serving as a foundation upon which the new structure is formed.

The alveolar process, being built or formed upon the body of the maxillary bones, conforms to them in outline and describes the same curves. In depth it corresponds to the length of the roots of the teeth, while in width it is sufficient to envelop all of that portion of the teeth located below the gum line. It gradually increases in width as it approaches the body of the bone upon which it rests.

It consists of an outer and inner plate united at intervals by septa, thus forming the alveoli for the accommodation of the roots of the teeth. In structure, the process is not compact, but open and spongy, somewhat resembling the cancellated structure of the diploë of the bones of the cranium or the inner portion of the shafts of the long bones. Its outer or cortical layer is denser and harder than the inner portion. Its cellular structure, while giving it sufficient firmness to support the teeth in their positions, affords opportunity for the lodgment and passage of the vessels of nutrition with which it is so bountifully supplied, and allows it to yield or bend in many cases when pressed upon by the moving tooth or teeth. This is especially true of the outer plate of the superior arch.

Owing to its peculiar structure and its great vascularity, the alveolus is readily resorbed under the stimulus of pressure and again readily reproduced behind the moving teeth.

THE TEETH.

Of the teeth themselves, but little need be said. The student is familiar with their number, shape, position and structure. Being the hardest structures of the human body, the application of any force necessary to move them will not injuriously affect them so far as their hard tissues are concerned.

A mechanical difficulty associated with their moving consists in the fact that their crowns are round and smooth, thus making it somewhat difficult to apply force at a given point. This difficulty, however, has been overcome by the introduction of the Magill band.

In considering the moving of teeth, the fact must not be

overlooked that while the crown may be moved considerably, the movement becomes less and less along the line of the root, so that the apex is moved but little. This is due to the fact that force can only be applied to the crown, while the apex remains almost a fixed point or fulcrum. In the movement of a tooth, therefore, whether inward or outward, forward or backward, the crown describes the arc of a circle, the centre of which is near the apex of the root.

Teeth with single and short roots can be moved more readily than those with long and many roots, for the reason that in the former case there will be less resistance to be overcome.

THE PULP.

The pulp is the formative organ of the tooth, and after calcification is complete it remains as the principal source of nutrient supply for the dental tissues, especially the dentine.

It is composed of fibrous connective tissue, containing a delicate system of lymphatics together with numerous nerve filaments which enter through the apical foramen. Ramifications of minute blood-vessels are noticeable throughout its whole extent, giving color to the organ and constituting its vascular system.

It bears an important relation to the teeth in their movement, since it may be readily devitalized through imprudence or lack of care. Before calcification of the teeth has been completed the apical foramen is large and easily accommodates the pulp where it enters the tooth. After calcification is complete the apical foramen is small, and the pulp at this point is in consequence greatly reduced in size. In the movement of the teeth there is often a slight mechanical constriction of the pulp at the apex due to the tipping of the tooth in moving. If the movement be rapid in teeth fully calcified (after the sixteenth or eighteenth year) this constriction may be so great as to cause death of the pulp through strangulation. Before complete calcification this is not likely to occur, from the fact that when the foramen is large the pulp has more space for its accommodation.

In the movement of a tooth in the direction of its length the pulp may also become devitalized through excessive stretching. This has occurred at times in drawing down into line a tooth that has been retarded in eruption. In all such cases care must be exercised, and the movement be conducted slowly.

THE PERICEMENTUM.

The pericementum or peridental membrane is that tissue which envelops the root of the tooth and fills the space intervening between it and the alveolar wall. It is a tough, strong membrane, composed mainly of fibrous connective tissue, permeated with blood-vessels and nerve fibres, and containing traces of a lymphatic system.

It is strongly adherent to the alveolar wall of the socket on the one hand, and to the cementum of the tooth on the other, its adherence being due to the extension of its fibres into both the bone and the cementum. These fibres, according to Prof. Black,* " are wholly of the white or inelastic connective tissue variety," and the apparent elasticity of the membrane is due to the passage of most of the fibres from cementum to wall in an oblique direction, in such a way as to " swing the tooth in its socket."

This membrane is the formative organ of the cementum of the tooth, and also assists in building the walls of the alveoli.

The cells concerned in the building of the bony walls are known as osteoblasts, and those forming the cementum are designated cementoblasts. After these cells have performed their normal function, they become encapsuled and form part of the tissue they were instrumental in building.

When re-formation of tissue is demanded, as in the thickening of the alveolar wall, or in increasing the normal

---

* Dental Review, vol. I., p. 240.

amount of cementum at various points under certain conditions, new cells are originated to perform the work. In the moving of a tooth the activity of these new cells is at once manifested in the formation of alveolar tissue to fill the space caused by the advancing tooth.

Beside these cells of construction and repair, the pericementum also contains cells that might well be called *cells of destruction*. They are the osteoclasts or cementoclasts, and their function is to break down or absorb the cemental or osseous tissues when nature calls for such action.

In the correction of irregularities these cells perform valuable service in removing bony tissue in front of the moving tooth.

The pericementum is thickest in childhood, when the sockets or alveoli are of necessity considerably larger than the roots of the teeth which they contain. With advancing age both cementum and the alveolar walls are increased in thickness by slow but continuous growth until the pericementum is greatly reduced in thickness, and in consequence the diameter of the roots more nearly approximates that of the alveoli or sockets.

The pericementum possesses a variety of function not often met with in any single tissue of the human system.

It retains the tooth in its socket and acts as a cushion to prevent injury to the adjoining bony structures from hard and violent concussions to which the teeth are sometimes subjected.

It affords accommodation for numerous blood-vessels which supply both the teeth and alveolar tissue with nutrient material, and for the branches of nerves which constitute it the sensory organ of the tooth, so far as tactual impress is concerned.

It is the organ of construction and repair of both cementum and bone, and is also, on occasion, the organ of destruction of either or both of these tissues.

PHYSIOLOGICAL ACTION IN THE MOVEMENT OF TEETH.

In the movement of teeth, one or both of two changes take place. One is the absorption of alveolar tissue on the advancing side of the tooth, and its reformation on the opposite side, and the other is a bending of a portion or the whole of the alveolar process surrounding the moving teeth. The former is the principal action that takes place in the moving of a single tooth, and the latter the chief one involved in the movement of a number of teeth in the same region, as in the widening or expansion of the arch.

When force is exerted upon a single tooth for the purpose of moving it, the first effect produced is the compression of the pericementum between the tooth and alveolar wall on the advancing side, and the stretching of the same membrane on the opposite side. In the compression of the membrane the blood supply is partly cut off, and the nerves, by their irritation, create a sensation of pain, which is soon obliterated by the semi-paralysis brought about by continued pressure. At the same time this irritation stimulates and hastens the development of the osteoclasts which at once begin the work of breaking down and absorbing that portion of the socket pressed upon.

Bony tissue being thus removed, accommodation is made for the advancement of the tooth, which at once takes place. Under continued pressure this action is renewed again and again until the tooth has reached its intended position. While this is taking place on the advancing side, quite an opposite condition prevails on the side from which advancement has taken place. There the fibrous tissue of the pericementum has been subjected to extreme tension, greater room has been provided for the accommodation of the nutrient vessels, and osteoblasts have been developed for the formation of bony material to add to the alveolar wall and thus close the space caused by the movement of the tooth. While these processes of absorption and reproduction on opposite sides of the tooth have been going on coincidently, their results have been very unequal, for the absorption of bone is a far more rapid process than its formation.

During the entire time of moving, and for a long time afterward, the tension of the pericementum on the free side of the tooth is kept up to such an extent that were the force of pressure or retention removed, the tooth would at once be drawn partly back into the space created by its movement.

The tendency is only finally overcome after the deposit of ossific matter in the alveolar socket has been sufficient to allow the pericementum to resume its normal thickness on that side of the tooth, when, by virtue of the removal of the tension and the support of the new bony tissue, the backward movement of the tooth will no longer be possible.

While this process of reparative construction has been going on, the structures about the opposite side of the tooth have been adjusting themselves to the new condition. The pressure upon the tooth having ceased, no more bone is absorbed; any injury inflicted upon the pericementum by its continued compression is repaired; the nerves and bloodvessels resume their normal functions, and the tooth in its new position becomes a far more useful member of the dental organism than it had been.

When a number of adjacent teeth are moved in the same direction, especially outward in the superior arch, or outward in the anterior portion of the lower arch, and inward in the posterior portion of the same jaw, a distinct bending of the alveolar process takes place. This is evidenced by the rapidity with which the movement takes place (which is far greater than the slow process of absorption would admit of), and also by the fact that after the movement is completed the process is not perceptibly thinner on the advancing side or thicker on the retreating side of the teeth than it was in the beginning. This flexibility of the process is due to its incomplete calcification at the age when operations for irregularity are usually undertaken. That the alveolar plate pressed upon should yield is more easily comprehended than that the opposite one should follow, but the septa uniting the two being similar in semi-cartilaginous strength and elasticity probably draw the one plate after the other as the movement takes place.

This flexibility of the alveolar process early in life is an important aid in the movement of teeth in that it both hastens and simplifies the operation, but it should not be presumed upon or too great advantage taken of it, for if an attempt be made to carry it beyond normal limits, the maxillary bones will be liable to separate along the sutural line, and the central incisors be spread apart in consequence.

PART II.

CHAPTER I.

MATERIALS AND METHODS.

EXAMINATION OF THE MOUTH.

When a case of irregularity presents for treatment, the first requirement is a careful examination of the mouth and teeth.

In conducting this examination it is necessary to note the position of the teeth, their relation to one another, their occlusion with those of the opposite jaw, the relative size and shape of both arches, the size, character and condition of the teeth, the age and general health of the patient, the harmony or inharmony of the features and the facial expression.

A careful consideration of all these points will enable us to decide:—

1st. What is desirable.

2nd. Whether it can be done.

3rd. If possible, how it can best be accomplished.

After this preliminary examination, our opinion of the case should be given the patient or parent accompanied by a plain statement of the difficulties of the case, if such exist, the probable time that will be required for correction, and an approximate estimate of the cost. To avoid any possible misapprehension the patient should also be informed that the appliances will cause some annoyance and possibly some pain, and that patience, endurance and perseverance will be necessary on his or her part to enable us to accomplish a satisfactory result.

It should also be mutually understood that the parent or patient shall assist in the furtherance of the work by seeing that the appliances are faithfully worn, that all the instructions are carried out, and that the patient shall punctually meet all appointments that may be made.

Should the prognosis of the case prove satisfactory and all of the above conditions be agreed to, we may at once proceed with the treatment.

IMPRESSION AND ARTICULATION.

The first step will be to take impressions of the upper and lower teeth from which to secure models for the further and more exact study of the case.

These impressions should be taken with some material that will receive a sharp imprint and not materially change its shape in removal from the mouth. Either Plaster of Paris or Modelling Composition (Stent's or Godiva) will give satisfactory results, but as the former can only be removed from the mouth by being broken into many pieces the latter is generally preferred. In selecting the impression cups, those known as flat-bottom cups should be chosen, on account of the better accommodation they afford for the crowns of the teeth. The cups should in all cases be large and deep enough to allow for a sufficient quantity of the material along the outer rim to enable a perfect impression to be taken of the labial and buccal surfaces of the teeth, and as much of the gum above them as possible. Figs. 11 and 12 represent cups of this character with high sides, devised by Dr. Angle for the taking of plaster impressions, but they answer quite as well for Modelling Compound.

A proper quantity of the composition having been softened by dry heat or in hot water, it is placed and properly shaped in the previously warmed cup and rapidly introduced into the mouth.

In taking an impression of the upper jaw the mouth should be kept well open so that the teeth may not come in contact with the material before the proper time and thus mar the surface. When the cup with its contents has been

FIGS. 11 and 12.—Angle's Impression Cups for Irregularities.

placed as far back as necessary, and immediately beneath the teeth, it should be brought up into position with a straight and steady movement. Once there, it should be firmly held while a finger is introduced to force forward into position the portion of material that has escaped at the rear of the cup, after which all that portion along the outer rim should be pressed against the teeth and gums from molar to molar.

In this position it must be held until it has become so hard that a finger nail will scarcely indent it, when it should be carefully removed. The hardening is best hastened by a stream of cold water from a syringe, or by the renewed application to the cup of small sponges or napkins dipped in ice water as suggested by Prof. Newkirk.

In taking an impression of the lower jaw the same general method is followed, and after the cup is in position all the surplus material around both the outer and inner rims should be pressed into place with the finger.

The models obtained from impressions taken in this manner will be sufficiently accurate to give us a good representation of both the buccal and lingual surfaces of the teeth, so necessary to a proper study of the case.

Impressions taken in plaster are the most accurate in detail, but the compound gives us all the accuracy we need in models for regulating.

During the same sitting at which the impressions are taken, the manner in which the teeth occlude should be observed and recorded, so as to enable us to place the models in proper relation while being attached to the articulator. This will dispense with the necessity for taking a bite.

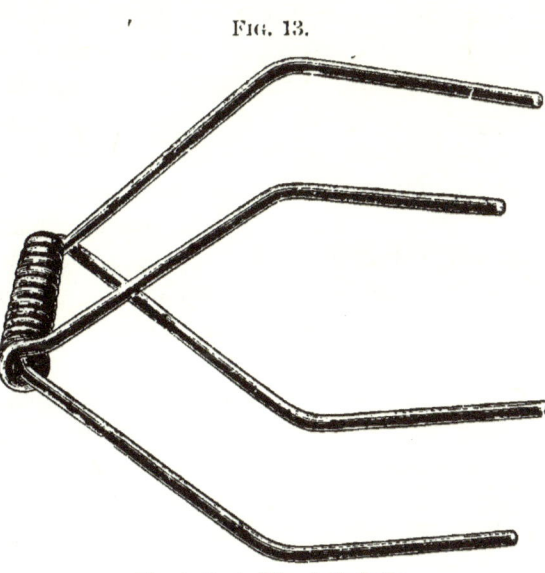

Fig. 13.

The Author's Wire Articulator.

An inexpensive and excellent articulator, Fig. 13, for the mounting of models of irregularity, is made from brass wire. The upper arms and coil are one continuous piece, while the lower arms are formed by passing another piece of the wire through the coil and bending to shape.

The articulator is so slender in outline that after the models are atttached to it the occlusion of the inner cusps of the teeth may be as readily examined as that of the outer ones.

With the models properly mounted on the articulator, our second and more deliberate study of the case may be carried forward at our leisure.

At the first or personal examination of the case, we are supposed to have decided upon the advisability of an attempt at correction, and also upon the general plan we purpose pursuing. By the study of the articulated models we will be enabled to decide upon the details of the work and the kind of appliance that should be used. Both studies are necessary, for with the patient in the chair we cannot take the time to map out the proposed work in detail, while an examination of the models alone will leave us without a knowledge of many important characteristics of the case that can only be gained from a personal examination.

Articulated models, made and mounted as described, are most important not only for purposes of present study, but also for comparison as the work progresses. Inasmuch as they represent the exact condition of the case at the beginning, we have in them a means of ascertaining what advancement has been made at any stage of the operation, whether the different movements are proceeding satisfactorily, and finally, when the operation is completed, of observing just how much change has been effected. An ocular comparison is of some value, but a mathematical one made with calipers and rule is far more exact and satisfactory.

STUDY OF CASE FROM ARTICULATED MODELS.

The study of the case may be either a simple or difficult one, according to the conditions and requirements involved. Thus, the movement of a single tooth will only involve the consideration of providing accommodation for it in the arch and the manner of applying force to bring it into position, whereas when a number of teeth in different locations are to be moved, each perhaps requiring a different form of movement, we will have to decide whether we can and should produce all of these movements with one appliance at one time, or whether it would be best to produce each movement separately and possibly with different appliances. If the latter, we will have to determine which should be accomplished first, which next, and so on.

For instance, where the entire upper arch is to be expanded to make room for outstanding cuspids, we will have three different operations to perform; the side teeth must be moved laterally, the anterior ones forward and the cuspids inward into line. To produce all of these movements at the same time with one appliance would be impossible from the nature of the case, therefore they will have to be performed separately, and usually in the order in which they have been named. In attempting to produce many movements with one appliance we often defeat our object, although occasionally, where the movements to be produced are of opposite character, we may advantageously play one against the other.

Where they are of the same character, or nearly so, too much should not be attempted at one time, for the loosening of many teeth will be liable to make our anchorage unstable, in which case we would have to suspend all operations until some of the teeth again became firm.

Having decided upon the order in which the movements should take place, we have two other important points to determine.

Amount of power required.—This will be determined largely by the age of the patient and the character of the teeth and process. As previously stated, early in life, before the process has become fully calcified, the teeth can be moved more rapidly than at a later period, and less power will be required to accomplish it; so also, in patients of the same age, the teeth of one will be more readily moved than those of the other. This is due both to the relative length of the roots and the resistance of the alveolar walls, and as we cannot judge of the lengths of the roots from the appearance of the crowns alone, we have to form our opinion in the matter from the general conditions.

Observation has shown that teeth with large crowns, situated in large and firm-looking jaws, usually have long roots; whereas, smaller teeth, associated with thin and more delicate processes, have shorter roots.

Therefore, considering the age of the patient and the appearance of the teeth and processes, we can at least decide whether the amount of force to be applied should be great or little.

Manner of applying power.—Among the many appliances or substances for yielding power in the moving of teeth, the practitioner has a range of choice from the screw with its directness and power, to the silk ligature with its gentle traction.

Between these two extremes we have materials that will yield us force in any desired degree. Selecting the one which seems best suited to the case, we must next decide upon the most advantageous manner of using or applying it.

There are two general methods of securing the power-producing appliances in the mouth. One is the use of a plate of some kind to which attachments can be made, and the other is the plan of attaching the appliances to the natural teeth in such a way as to dispense with the wearing of a plate.

In certain methods of regulating, such as Angle's, Jackson's and Patrick's, no plate is used; while in others, such as Coffin's, a plate is invariably used for attachment and security. Farrar advocates the use of a plate only in exceptional cases. Each manner has its advantages and disadvantages. In the use of a plate, we have as advantages:—

Its convenience and adaptability.—Covering a large surface, it affords opportunity for the attachment of the immediate power-yielding appliance in any position and at any angle, and permits the same to be altered or changed with very little trouble. It also protects the soft tissues from any possible injury which might result from the slipping or impingement of other appliances upon them. Indeed, in many cases, a plain rubber or metal plate covering the roof of the mouth and not having any appliances attached to it, is used simply for the protection of the gums during the operation of regulating.

Its distribution of the power of resistance.—Touching all or nearly all of the teeth not being operated upon, it compels each one to bear its part in offering resistance to the power used for the movement of certain teeth, and in this way brings more teeth into use as points of resistance than can possibly be done by any other method.

Its simplicity of construction and the facility it affords for adjustment and alteration.

The disadvantages pertaining to the employment of a plate as an aid in regulating, are:—

Its uncleanliness.—Inasmuch as a plate comes in contact with so much tooth surface at the necks and elsewhere, it offers special opportunity for the accumulation of debris. In plates that are removable by the patient, this may be largely avoided by frequent cleansing, but observation has shown that the majority of patients are either so careless or indifferent in regard to the matter, that a clean regulating plate is seldom seen. In plates so constructed or arranged that only the dentist can remove them, the uncleanliness of the plate and consequent danger of injury to the teeth is greatly increased

The frequent appointments necessary.—In the class of plates last alluded to, it is absolutely important that they be removed and cleansed at least once in every forty-eight hours. This requires such frequent visits on the part of the patient and the expenditure of so much valuable time on the part of the operator, as to constitute a serious objection to the use of such plates where they can at all be dispensed with.

When plates are not used, appliances are usually attached directly to certain teeth which serve as anchorages. Such attachment is generally secured by means of bands or collars encircling the teeth and cemented to them; or, in other cases, by having the bands simply passed around the teeth of attachment and drawn tight by means of screws or clamps.

The advantages of appliances attached to the teeth in this way are:—

1st. The leaving of the roof of the mouth uncovered, thus affording more room for the movements of the tongue.

2nd. Their greater cleanliness, because they touch the teeth at few points, and thus furnish good opportunity for thorough cleansing with the brush.

3rd. Not needing to be removed, fewer visits to the dentist are necessary, thus effecting a great saving in time and labor.

The only objection that could be raised against this manner of attachment is, that fewer teeth are brought into service in anchoring the appliance, but this objection may be largely overcome by making such extensions or additions to a band as to cause teeth adjoining the banded ones to bear their part in offering resistance. Two extensions of this character, as used in the author's practice, may be seen in cuts 14 and 15.

In Fig. 14 a bicuspid is banded, and to the band on the buccal side is soldered a strip of platinized gold long enough

MATERIALS AND METHODS. 73

to reach to and rest upon the adjoining teeth, causing them to bear their part in affording the needed resistance. We thus get the resistance of three teeth with the use of a single band.

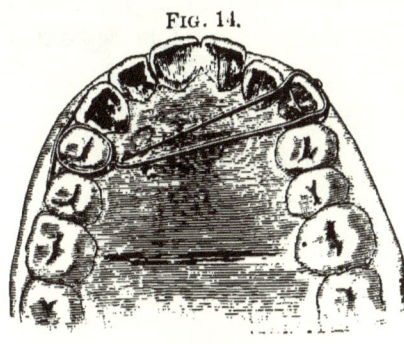

Fig. 14.

In Fig. 15 an extension strip of gold is soldered to a bicuspid band, in order to obtain the additional resistance of the adjoining molar.

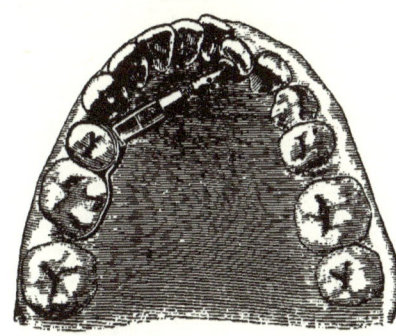

Fig. 15.

Dr. Angle recommends the banding of two adjoining teeth and having these bands united before being placed in position, as shown in Fig. 16. He claims that by this method the resistance is greatly increased, for the anchor teeth to move at all would have to move bodily forward in an upright position instead of tipping. So much resistance to this form of movement would be offered by the mass of alveolar tissue involved, as to make it almost impossible of accomplishment.

Fig. 16.

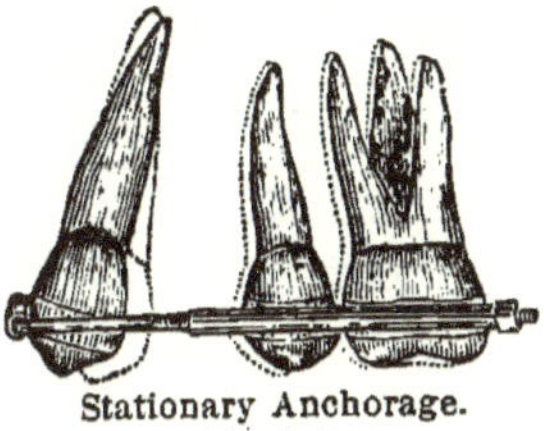

Stationary Anchorage.
(Angle.)

When bands are placed around teeth and secured by some mechanical device, they never can fit the teeth so accurately as to avoid spaces for the accumulation of food and saliva. The fermentation of the particles of food, and the acidity of the saliva in a state of rest, will soon injuriously affect even good tooth structure.

This can only be prevented by the employment of some material that will perfectly fill the space between the band and the tooth, therefore all bands passing around and encircling the teeth, in order to be harmless, should be cemented in place with phosphate of zinc.

CHAPTER II.
APPLIANCES.

MATERIALS AND THEIR USES.

During the study of the case, after we have decided upon the amount and kind of power we wish to apply in order to produce the desired movements, we will have to consider the different materials at our disposal in order that we may select from them the ones best suited to our purpose for the case in hand.

PLATINUM AND ITS ALLOYS.

Platinum on account of its tastelessness, its non-oxidability and its harmonious color, should constitute it one of the best metals for use in the mouth. Its extreme pliability and softness, however, greatly limit its usefulness, so that it can be used only where these latter qualities do not interfere with its employment.

It is chiefly used in the construction of bands that are to be cemented to the teeth to serve as anchorage for appliances or to form parts of retaining fixtures.

In combination with other metals, in the form of alloys, its greatest usefulness is developed.

IRIDIO-PLATINUM.

This alloy, combining the color and purity of platinum with the hardness and stiffness of iridium, is useful for bands, bars and wires, in connection with regulating appliances where platinum alone would not be available on account of its softness.

PLATINIZED GOLD.

Gold in a pure state, or alloyed with silver or copper, does not possess the stiffness necessary for its use in the form of bars, springs or accessories, where great resistance or elasticity is requisite, but when alloyed with about five per cent. of platinum it attains a degree of elasticity second only to steel. In this form it is one of our most useful materials for even the heat of soldering does not rob it of its elastic quality.

This alloy of gold can be purchased in the dental depots in plate of any thickness and in wire of any form or size. When used for the construction of screws or supports, its stiffness is the quality taken advantage of, while in the form of levers or bows its elasticity constitutes its chief excellence.

PLATINIZED SILVER.

This alloy, though long and favorably known in England, has never been extensively used in America. It is prepared for the market in the form of plate and wire of every gauge. In the form of plate it is largely used abroad as a base for artificial dentures, especially small partial pieces, while the wire is used as a support for the Ash tube-teeth and other purposes.

The alloy is composed of one part of platinum to two of silver. Its stiffness and elasticity is but little inferior to platinized gold, while its cost is about one-half that of gold. It can be rolled, bent or fashioned in any form and may be soldered with the highest grades of gold solder.

In the form of wire the author has found it very useful in the construction of bows for the attachment of rubber bands or ligatures to draw teeth in any direction, and for parts of retaining appliances where inconspicuousness is desirable.

Its non-oxidability is also a feature of considerable value.

GERMAN SILVER.

This improperly named alloy, composed of copper, zinc and nickel, is frequently employed in the construction of regulating appliances, on account of its stiffness and inexpensiveness. While it may be regarded as a base compound, its baseness is of so high a grade that it may be used without fear of harm to the soft tissues or the general system. Prof's. Angle and Matteson use it very largely in the construction of their appliances, and the author has made frequent use of it without ever noticing any deleterious effects. Its valuable qualities are many, and the ease with which it may be electro-gilded furnishes us with a means of improving its appearance.

GOLD.

Gold, in its non-elastic condition, has been and probably always will be one of the most useful of the metals for the construction of parts of regulating appliances. Its softness, adaptability and strength are all qualities of the greatest value and render it serviceable in numberless ways. To preserve its purity, and as far as possible to prevent oxidation, it should never be used of a carat less than 20 or 22.

STEEL.

This metal has the same desirable qualities of firmness and elasticity that are found in platinized gold, and possesses them in a higher degree, so that it is used in preference to the former metal where greater power is needed.

There are two disadvantages, however, connected with its use:—one is, that it cannot be highly heated (as in soldering) without destroying its temper; and the other, that it oxidizes so readily when in contact with the fluids of the mouth. This latter objection is largely overcome by electro-plating it with gold, a full description of which method will be found in the latter part of this volume. It is used principally in the construction of jack-, and other screws and as wire in the form of bows, levers and springs.

VULCANITE.

Soon after the introduction of vulcanite as a base for artificial teeth, its qualities of adaptability, strength and elasticity were recognized and utilized in the construction of appliances for regulating. By its use we secure advantages that could not be so readily gained from other substances.

Used either to produce pressure by its own elasticity, or as a medium for the attachment of other power-producing appliances, it has been one of the most commonly employed materials for the construction of regulating appliances.

COMPRESSED WOOD.

The use of this substance is very old. Before the introduction of either soft or vulcanized rubber, the quality of the expansion of compressed wood under moisture was employed in lieu of elasticity.

It was chiefly used in the form of small sections placed between a silver or gold plate and the teeth to be moved, a suitable slot or socket for its retention having been formed in the plate.

In this way it is no longer used, other materials possessing superior qualities having superceded it.

The author occasionally finds great advantage from the use of compressed wood in the separation of teeth for the accommodation of some malposed tooth, where the existing space, though not sufficient, is still too great to admit of the use of elastic rubber.

In such cases it is his custom to cut a cross-section from some compressible wood, such as cotton-wood, a little larger than the space it is to occupy. This is compressed in the direction of the length of the fibre by means of a hammer, after which it is notched at each end to fit the convex surfaces of the teeth to be moved. Upon being placed in position its expansion by the absorption of the fluids of the mouth will quickly cause the movement of the teeth. In the course of its expansion it adapts itself accurately to the tooth surfaces and thus does not become dislodged or slip from its position.

SEA-TANGLE.

This is one of the newer substances introduced into the list of materials that are of service in regulating. The idea of its use was borrowed from the medical fraternity, who first employed it for distention of the cervex uteri. It is a variety of sea-weed botanically known as *laminaria*, that has been robbed of its moisture and compressed until its density is about equal to horn. For medical use it comes in the form of a cylindrical tent about one-fourth of an inch in thickness and two inches in length.

This is the only form in which it has been placed upon the market. In the presence of moisture it rapidly expands from two to three diameters. As it expands only in the direction of its width, sections from it must be so cut and shaped as to take advantage of this lateral enlargement.

In regulating it may be employed in place of compressed wood, and like it is used to produce pressure between the unyielding plate and the tooth to be moved. A place for it is readily provided by cutting a hole or socket in the rubber plate at the desired point.

Its advantage over rubber or wood lies in its greater expansive properties and the ease with which it can be secured in place. A piece of suitable size can be placed in position and the plate properly secured in the mouth before expansion begins.

ELASTIC RUBBER.

The resilience of elastic rubber was early recognized as a valuable property that might be used to advantage in producing traction upon teeth to be moved. It was first used in the form of strips attached at either end by ligature, but since the introduction of rubber tubing, rings or bands cut from the same have been employed instead. Their first employment has been credited to Dr. E. G. Tucker, of Boston, about the year 1846.

These sections, cut from the smaller sizes of French rubber tubing, are now in almost universal use in connection with other appliances for regulating, and their value has been greatly enhanced since the Magill band has furnished a better means for their attachment.

Their power, though great, is limited, for they cannot exert so great a force as the metals, but their wide range of applicability and the persistence of their power places them among the most valuable adjuncts of regulating devices.

In use, their tendency to slip off the tooth or up under the gum (which constitutes the chief objection to their employment) must be guarded against by so securing them that change of position will be impossible. They should never be permitted to rest upon or touch the soft tissues at any point.

SILK LIGATURES.

The contraction of silk, linen or cotton thread in contact with moisture, enables us to make use of it where the gentlest tractile power is desired. Most frequently it is employed simply as a ligature in attaching some appliance to the teeth, but it has often been used to advantage in cases where teeth were to be moved slowly and a very short distance. Prof. Peirce employs it in this way for the moving of certain single-rooted teeth, as described in Part III. Its gentle power, together with its safety and simplicity, will often prove the very qualities we desire in certain simple operations.

CHINA-GRASS LINE.

This material has been extensively used in New England for ligatures in regulating, being preferred for that purpose to silk or cotton.

It is the *Boehmeria nivea* of botanists, and more commonly known as Ramie or Rhea fibre, and is the material from which China-grass cloths are manufactured. It is stiff enough to be threaded with a pair of tweezers between the teeth at their necks, thus avoiding the pain of forcing a ligature between them when tender.

It is non-elastic, but shrinks greatly without softening when moist, thus exerting considerable traction without producing pain.

QUALITIES AN APPLIANCE SHOULD POSSESS.

In selecting a form of appliance from among the many that have been devised by writers and workers in this field of practice, or in devising one to suit the demands of the case under consideration, it will be well to consider and bear in mind the qualities any appliance should possess in order to render it most effective.

The following are among the most important of such qualities:—

Efficiency.—The first requirement of any device is, that it shall be able to do the work expected of it. All appliances are, of course, devised with this end in view, but the attainment of it is often not as simple a matter as might at first appear. Almost every case has associated with it so many features and peculiarities claiming consideration, that even with the greatest care and thought we often fail to apprehend or grasp each individual complication. Some, indeed, are so little apparent that they can scarcely be recognized in advance.

For this reason even the most experienced practitioners will at times devise an appliance which, though seemingly meeting all the requirements, will, when brought to a practical test, fail to accomplish the end desired. It will then have to be altered, or perhaps discarded in favor of some other fixture more perfectly adapted to the requirements of the case.

An appliance that will not yield the results we desire, or which yields them in an imperfect manner, should in all cases be superseded by another.

Simplicity.—A complicated device is in nearly all cases less efficient than a simple one. Simplicity is a cardinal virtue in all matters of construction, and through lack of it about seventy-five per cent. of the patents granted in this country prove unprofitable.

Far greater mechanical ingenuity is displayed in an effective simple device than in a complicated one.

Rapidity of action.—In order to lessen the discomfort of the patient and to conserve the time of both patient and operator, a regulating appliance should be as rapid in its action as is consistent with physiological conditions. Too rapid action may cause suffering to the patient and possibly bring about deleterious results, while too slow action will prolong the treatment unnecessarily and possibly cause the patient to become disheartened and abandon the treatment.

Between these two extremes there is a mean in which the best results are accomplished.

All regulating appliances are at best a source of some discomfort to the patient. A foreign body in the mouth, occupying a certain amount of space and thereby interfering more or less with natural functions, cannot fail to be objectionable. In order, therefore, to lessen his discomfort as much as possible, we should try to devise appliances that will occupy no more space than is necessary and also have them free from all rough projections. Very little is required to cause abrasion of or injury to the soft tissues of the oral cavity, and when once caused such lesions are the source of much pain.

Least interference with speech and mastication.—Most patients apply to us for correction of irregularity at a time when their education is in progress. Their lessons must be recited, and their enunciation must be distinct enough to be understood by the teacher. With a large and cumbersome appliance in the mouth it would prove very difficult for them to speak distinctly, and they would thus be placed at a disadvantage.

They are also in their growing age when the body needs an abundance of nutritious food to supply the demands of the various tissues. If mastication be insufficient through imperfect occlusion or through tenderness of the teeth caused by a bulky fixture, nutrition will be inadequate to the needs of the system.

Such conditions can and ought to be avoided by a properly constructed appliance.

Cleanliness.—The cleanliness of any appliance will depend both upon the method of its construction and the care that is taken of it. If it be removable so that the patient can take it out, cleanse and reinsert it, there ought to be no difficulty about its being kept clean. The patient should be instructed to remove it for cleansing at night, in the morning, and after each meal, at the same time giving the natural teeth a thorough brushing.

A good plan is to supply the patient with a brush, properly marked, to be kept in the office. When the patient appears and the appliance is removed, the operator should see that both plate and teeth are well cleansed in his presence. This one cleansing he will be sure of, though he may not be certain of the others. The same plan is pursued with plates or appliances that can only be removed by the operator. Where appliances are of such character that they seldom need to be disturbed, the patient should be taught to take a quantity of water in the mouth, and then using the lips and cheeks bellows-fashion, force the water through every interstice of the teeth and appliance to flush out accumulations. This should be done each time after eating as well as before retiring and after rising.

Most appliances can be worn a long time without injury to tooth substance, if they are properly constructed and kept scrupulously clean.

Without cleanliness, the teeth will soon be injured by the secretions and accumulations, and the breath of the patient, from the same cause, will become so offensive as to disgust all brought within its range.

Inconspicuousness.—Annoyance from wearing a conspicuous appliance is often added to the other ills which the patient is subjected to during the process of regulation. An appliance of this character, while often producing distortion of the lips, also attracts much attention and compels the wearer to make frequent answers to the same oft-repeated question.

Young persons attending school or entering society are

naturally very sensitive to the ill-appearance of any conspicuous device. Whenever the same result can be accomplished by a concealed fixture as by an exposed one, it is better to adopt the former; but where a better or more satisfactory result can be obtained by the use of a more prominent fixture, appearance will, of course, have to be subordinated to utility.

Stability.—The quality of stability has previously been spoken of, but its real practical importance cannot be too strongly insisted upon. It is a *sine qua non* in orthodontic practice. With it, we have a reasonable certainty of results; without it, all is uncertainty.

In some cases, as where most or all of the superior teeth are to be drawn backward, we have apparently no point for proper anchorage. Stability or fixedness of position for an appliance, in such cases, not being obtainable within the mouth, some fixture can be devised which will have its point of resistance outside, as on the back of the head.

This plan of securing resistance outside of the mouth, has been adopted thus far only in a few exceptional cases, but it is hoped that its advantage and importance will lead to its more frequent employment in the future.

Freedom from injury to tooth substance.—By this we do not mean chemical injury, for that has already been treated of, but we refer to mechanical injury. Any sharp, hard point or roughness of a metallic appliance, will be likely to scratch and mar the surface of enamel and thus prepare the way for future decay.

Steel jack-screws of any form, when placed directly against the teeth of anchorage and those to be moved, are liable to work injury to tooth structure. For this reason there should always be interposed between the teeth and screw some material that is non-injurious to the tooth. Besides protecting the teeth, such substance will also serve to give greater security to the screw.

To obtain this same fixedness for the point of a fish-tail jack-screw, or other appliance, some operators have been in the habit of drilling a hole or depression in the tooth to be moved. It is hoped that the introduction of the Magill band has caused the abandonment of this practice, which at best was only justifiable in exceptional cases and in self-cleansing localities.

RETAINING APPLIANCES.

The retention *in situ* of teeth that have been moved, for a time sufficiently long to allow them to become firm, is quite as important as the moving of them. As previously explained, teeth become firm in their new position by virtue of a deposit of ossific matter in the space created by their displacement. The formation and perfect ossification of this new material is only completed after a lapse of time varying with the age and constitution of the individual. Experience has proven that a less time than six months should never be allowed for it, while in persons of mature age or in those younger where many teeth have been involved, the time will sometimes have to be extended to a year or longer.

The natural tendency of a tooth to return to its former position, aided by the tension of the parts that have resisted its movement, will certainly move a tooth from its new position, unless the newly formed process has become thoroughly calcified, and is thus by its strength and density able to resist the opposing forces. Numberless failures to retain the good results of regulation are attributable to this cause alone.

In certain cases, as where a superior incisor has been occluding inside of the lower ones, or where a lower one has been biting outside of the upper ones, no retaining appliance will be required after they have been brought into proper position, because the natural occlusion of the jaws will prevent the corrected tooth from returning to its former position.

So also with the bicuspids and molars. Where malocclusion has forced them out of their true position, or kept them there, the correction of the occlusion will often tend to retain them in their normal positions without extraneous aid.

In all other cases, however, mechanical assistance will be necessary until the teeth have become firm. Where the arch or any portion of it has been enlarged, or where a number of teeth have been moved from within outward, the simplest and probably the best means of retaining them will be the wearing of a thin rubber or metal plate covering the palatal arch and nicely fitting each tooth at its neck. It may contain a vacuum-chamber or not, as preferred, but in many cases the use of one will greatly assist in keeping the plate in place. In addition to its use in preventing teeth from moving inward, the plate may often be advantageously modified by the addition of a gold hook or spur to keep rotated teeth in position, or to retain individual teeth that have been moved inward.

While rubber plates in some form, either by themselves or in combination with accessories, are frequently used for retaining corrected teeth, their use is, nevertheless, open to certain objections. All rubber plates used either for correction or retention, must be removed at frequent intervals for cleansing. The very necessity for their removal affords opportunity for the patient to remove them at other times, and possibly forget or willfully neglect to reinsert them for a longer or shorter period, thus causing delay in the reparative process.

Besides this, also, in the very act of removal and insertion the teeth are slightly moved in their sockets, and this will to a certain degree hinder the re-formation of tissue.

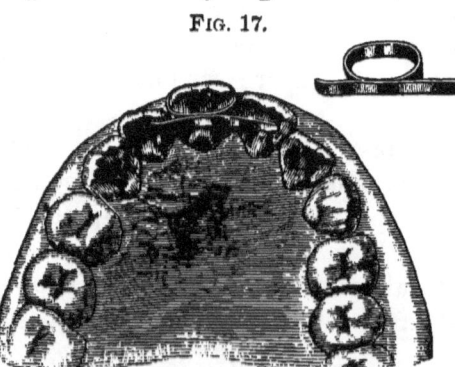

Fig. 17.

The Author's Band and Bar Retainer.

MATERIALS AND METHODS.   87

On account of these objectionable features the author has for many years avoided the use of rubber retaining plates, wherever he could do without them. As a substitute he was led to devise a number of little appliances of gold and platinum, occupying the least possible space, and firmly attached to the teeth for the required time. Fig. 17 shows one of these appliances in its simplest form. It consists of a platinum (Magill) band, freely fitted to the tooth, and having a gold bar or spur soldered to it to press or bear against one or more of the adjoining teeth. When properly adjusted, it is secured to the corrected tooth by means of phosphate of zinc.

As will readily be seen, its advantages consist in its small size, its slight contact with teeth other than the one upon which it is placed, its cleanliness, its fixedness and the firmness with which it holds the corrected tooth in place. The latter is its most important feature, for it is a well recognized fact in surgical practice that, other things being equal, reunion of bony tissue or new formation of the same will progress in rapidity proportionate to the stability of the parts.

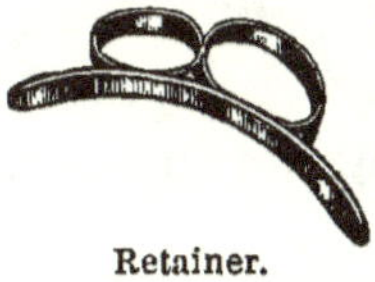

Fig. 18 shows a modification where two teeth are thus to be retained with the extension bar long enough to include more distant teeth. Fig. 19 represents two bands joined at their borders, for the retention of two teeth that have been rotated.

Still another modification is shown in Fig. 20. In this case the two bands on the cuspids are united by a thin gold or platinum wire passing along and conforming in outline to the labial surfaces of the inter-

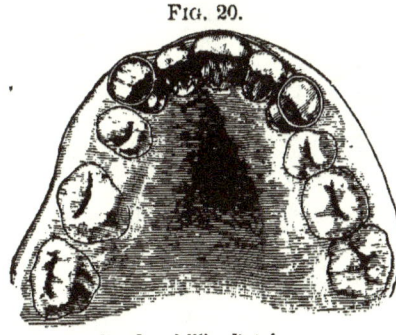

FIG. 18.
Retainer.

FIG. 19.
Retainer.

FIG. 20.
Band and Wire Retainer.

vening teeth. It was used to retain three incisor teeth that had been drawn inward.

Fig. 21 illustrates a retainer of nearly similar character for the lower incisors. In this case a band of gold takes the place of the wire on account of its greater stiffness.

Retaining appliances of this character cannot, of course, be used to advantage in all cases; but where they can they will be found to be most satisfactory.

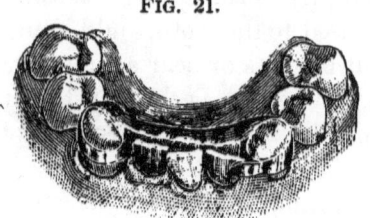

FIG. 21.

Prof. Angle uses a retaining appliance differing from the foregoing in having a tube soldered horizontally to the band that encircles the tooth. The tooth once in position a wire is passed through the tube and made to rest upon the adjoining teeth, after which a hole is drilled through both tube and wire and a short pin inserted to prevent the wire from shifting its position. See Fig. 22.

FIG. 22.
Angle's Retainer.

Another simple and ingenious device for retaining teeth after they have been moved, especially after rotation, was shown the author by Dr. H. A. Baker. It consists of a gold screw cemented into some conveniently located cavity in such a way that the protruding portion will rest against an adjoining tooth, and thus prevent the tooth

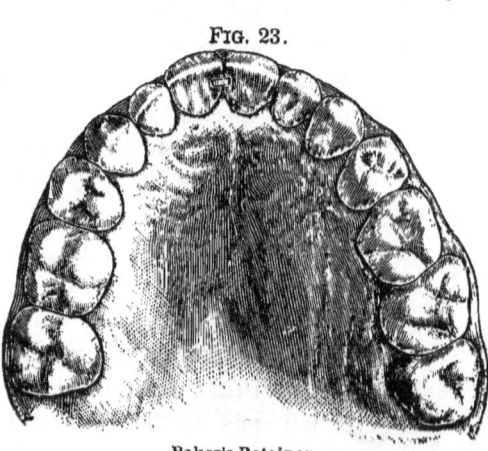

FIG. 23.
Baker's Retainer.

operated upon from changing its new position. Such device could, of course, only be used in rare and exceptional cases; but where applicable, it possesses the advantages of simplicity, inconspicuousness and efficiency. Fig. 23 represents a case in which a rotated incisor was thus retained.

A very simple appliance for holding teeth that have been drawn toward one another is shown in Fig. 24, and was devised and first used by Dr. C. S. Case. It consists of a silver or platinum wire passed over lugs or pins upon bands attached to the teeth to be retained. Floss silk or China-grass line, used in the same manner would answer instead of wire; but they would neither be as strong nor as cleanly. Dr. Case also uses the wire for exerting a gentle traction force where needed by soldering a piece of square metal tubing to it at about the middle of its length and turning this with a suitable instrument, thus twisting the wires and drawing the teeth together.

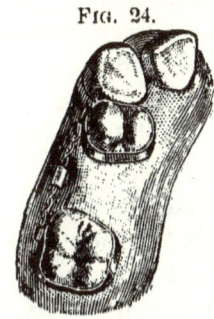

FIG. 24.

CHAPTER III.

CONSIDERATION OF METHODS.

FARRAR'S METHOD.

In 1876 Dr. J. N. Farrar began publishing a series of articles in the *Dental Cosmos*, descriptive of a method he had devised for the regulation of teeth. Reading and observation, he said, had satisfied him that the various plans suggested up to that time for the correction of irregularity, were lacking both in system and principle. He claimed that the performance of so important an operation as regulation should be based upon a correct knowledge of both mechanical and physiological law.

Experience had convinced him that the character of force applied to the teeth should be *positive*, and that it should be *intermittent*—a period of rest following a period of motion.

The best instrument for applying a force that is positive and may be intermittent, he said, was the screw in one of its various forms.

Experimenting with appliances constructed upon the screw principle, convinced him that this method of delivering force was not only positive and direct, but also that its range of applicability was so great that it might be used to the best advantage in nearly all cases of regulating. He claimed, also, that it was the only instrument whose force could be controlled at will and thus be made to exert power upon or retain in a state of repose the tooth or teeth operated upon.

This alternation of motion and rest was as important in changing the position of teeth as in other organs of the body, and was in strict accord with physiological law. In his experiments he found that intermittent force was productive of less pain to the patient than continuous force, and might be so skilfully applied as to prevent all pain.

Pain, he said, was an expression of a pathological condition, and by its avoidance we kept within the boundary separating the physiological from the pathological state. With screws of known pitch and number of threads, he found that he could move a tooth painlessly, and therefore safely, from $\frac{2}{240}$ to $\frac{1}{160}$ of an inch every twenty-four hours. His experiments led him to the following conclusions:—*

"*1st.* That in regulating teeth, the traction must be intermittent, and must not exceed certain fixed limits.

"*2d.* That while the system of moving teeth by elastic rubber apparatus is unscientific, leads to pain and inflammation, and is dangerous to the future usefulness of the teeth operated upon, a properly constructed metallic apparatus, operated by screws and nuts, produces happy results, without pain or nervous exhaustion.

"*3d.* That if teeth are moved through the gums and alveolar process about $\frac{1}{240}$ of an inch every morning, and the same in the evening, no pain or nervous exhaustion follows.

"*4th.* That while these tissues will allow an advancement of a tooth at this rate ($\frac{1}{240}$ of an inch), twice in twenty-four hours, the changes being physiological, yet, if a much greater pressure be made, the tissue changes will become pathological."

The above conclusions were epitomized by him into the following Law:—"In regulating teeth, the dividing line between the production of physiological and pathological changes in the tissues of the jaw is found to lie within a movement of the teeth acted upon, allowing a variation which will cover all cases, not exceeding $\frac{1}{240}$ or $\frac{1}{160}$ of an inch every twelve hours."

---

* *Dental Cosmos*, Vol. XVIII, p. 23.

His articles upon the subject may be found in the *Dental Cosmos*, extending from Vol. XVIII to XXIV.

Although the screw principle was the one which he principally used, and the only one which he considered scientifically and physiologically correct, he at times availed himself of the use of some of the continuous-force appliances, such as rubber bands,* silk or fibre ligatures,† and, for the attachment of appliances, vulcanite plates.‡

So far as the principles upon which Dr. Farrar's system is based are concerned, they have received but limited endorsement on the part of the profession, but the multiplicity and variety of his appliances and the ingenuity displayed in their devising, have commanded the admiration of all and been of great value to laborers in this field. Most of his appliances are original in design, beautiful in construction, and well calculated to perform the work intended; but in confining himself so largely to the use of one form of power-producing instrument, his apparatus is in many cases very elaborate and complicated. The same end could often be accomplished by much simpler means.

His appliances are so numerous that illustrations of many of them could not be introduced into a text book, nor could they well be selected from to illustrate his principles, but some of them may be found in Part III, where the practical treatment of various forms of irregularity is considered.

Dr. Farrar has recently published in book form a full elaboration of his views and methods, together with numerous illustrations of his appliances, to which the reader is referred.

PATRICK'S METHOD.

In 1882, Dr. Patrick brought forward his method of regulating. His appliance is made of gold, and designed to be

---

* *Cosmos*, Vol. XIX, p. 520.
† " " XXI, " 306.
‡ " " XXI, " 306.

attached directly to the teeth on presentation of patient, without the usual preliminaries of taking an impression and making a model.

The appliance with its appurtenances, all beautifully constructed and ready for use, may be purchased from the inventor or through the dental depots.

The essential parts consist of a bow-spring, adjustable anchor bands, and numerous devices for engaging with the teeth to be moved. Fig. 25 represents the appliance with many of the accessories in position. The bow-spring "A" consists of a half-round bar or wire of platinized gold, bent in horse-shoe form to approximately conform to the shape of the arch. " BB " are the adjustable loop-bands, made of thin gold plate, the free ends of which, on their palatine surfaces, are connected with a screw and fixed nut (C), for bringing the band in close contact with the tooth to which it is applied. On the buccal surfaces of these bands are soldered sections of half-round tubing, accurately fitting the bow-spring which plays through them. Outside of this tubing is soldered a nut threaded to receive the long buccal screw (D) intended to tighten the bow-spring after it is in position, or to take up the slack caused by the moving teeth. The head of this screw passes through and operates against a smooth nut soldered to a section of the tubing which is temporarily attached to the bow-spring at any point by means of an adjustable double wedge.

Of the accessory appliances shown, " E " is a hook intended to rest against the mesial or distal (by reversing) surfaces of a tooth intended to be moved in an anterior or posterior

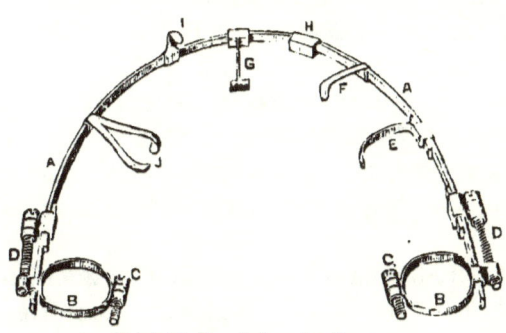

FIG. 25.

Patrick's Regulating Appliance.

direction. It is secured in the desired position by a wedge placed between the inner flat surface of the spring and the adjoining surface of the slide to which the hook is attached. The tooth is moved anteriorly or posteriorly by loosening the buccal screw on one side and tightening it on the other toward which the tooth is intended to be moved. " F " is a hook to catch over the cutting edges of incisors when it is desired to move them from within outward. When in position the tension of the bow-spring is increased from time to time by unscrewing the buccal screws. " H " is a slide or section of the half-round tubing, reinforced on its palatine surface by additional gold, and intended to be used as a stud to prevent one portion of the tooth from moving while the opposite one is being moved, as in rotating an incisor. It is also used to press against any tooth as a wedge in moving it inward.

" G " is a T-bar for producing double rotation of the incisors. " I " is a hook set vertically to engage with the cutting edge of an incisor, to prevent the bow-spring from slipping up toward the gum.

" J " is a bifurcated hook, to grasp a cuspid tooth intended to be moved outward.

Each of these appurtenances is soldered to a section of the half-round tubing, which allows it to be moved to any desired position on the bow-spring. When in position it is retained by means of the wedge already referred to.

As will readily be seen, the power obtained by this appliance consists partly in the elasticity of the bow-spring and partly in the direct action of the tightening screws.

The ingenuity displayed in the devising of this method is certainly very great, and the delicacy and accuracy of construction of the various parts all that could be desired. The combination of the principles of the spring and screw bring into play two of the most important powers available in regulation, and their correlation in this method is very happily brought about.

BYRNES' METHOD.

Dr. B. S. Byrnes has devised a method for regulating teeth by the use of narrow strips of fine gold variously shaped and bent to produce tension upon the malposed teeth. The method is an exceedingly novel and ingenious one, and while it could not be used to advantage in all cases, still contains elements of merit that will be of value to the practitioner. His power is derived from the elasticity of the metal, which is corrugated in such a manner as to develop this quality to the highest degree.

His bands are made from gold plate of 20 to 22 k. fineness, rolled very thin, and when greater power is needed the bands are doubled in thickness. He uses no plates, but anchors his appliances by means of bands to suitable teeth, situated at some distance from the ones to be moved.

* The method of application, in a general way, is as follows:—The fixed points having been determined upon, the tooth or teeth to be regulated are connected with them by means of a thin gold band. The band is manipulated so as to form it into a spring, or series of springs, so adjusted as to bear most powerfully on the misplaced tooth. Thus, supposing that a projecting superior central incisor is to be drawn inward to align properly with the remainder of the teeth in the arch, a continuous gold band embracing the first molars on both sides is fitted around the outside of the arch.

With a dull-pointed instrument, like a burnisher, the ribbon is then pressed into the interstices of the teeth over which it passes, thus forming it into a series of small springs. The incisor being the most prominent point will naturally be most affected by the pressure exerted by the springs, and in a short time it will be found to have moved away from the band, so that it is no longer affected by its tension. As soon as this occurs the apparatus is removed, the ribbon annealed and straightened, and a small portion,

---

* *Dental Cosmos*, Vol. XXVIII, pp. 278-284.

say a thirty-second to a sixteenth of an inch, as may be required, is cut out of it. The ends are then soldered and the appliance replaced upon the teeth, the connecting band being formed into a spring as before. Tension is thus kept up until the tooth has assumed the desired position. Sometimes the spring of the band may be advantageously supplemented by other aids, as the insertion of a rubber wedge (under the band) at points where a particular gain is desired.

Figs. 26, 27 and 28 illustrate the general appearance of the appliance in some of its forms.

Figs. 26 and 27 were used to draw in projecting incisors in the case of a young lady, aged 18. The movement was assisted by rubber wedges placed between the band and the labial surfaces of the teeth. "The connecting band was cut and shortened every other day, the patient having a sitting every day to allow the band to be sprung more as the teeth moved away from it."

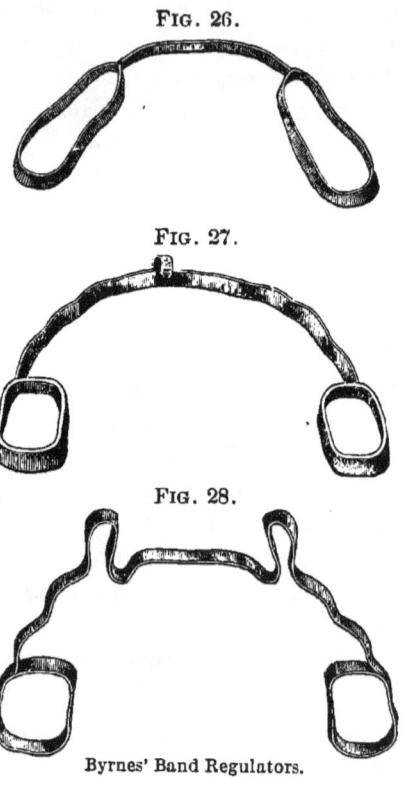

FIG. 26.

FIG. 27.

FIG. 28.

Byrnes' Band Regulators.

Fig. 28 represents the form of appliance used by Dr. Byrnes in drawing forward the lower incisors, and pressing back the cuspids at the same time. The band clasped the first molars of each side and passed around the cuspids and back of the incisors. By cutting and shortening the band from time to time as the teeth yielded to the pressure, the irregularity was easily and quickly corrected.

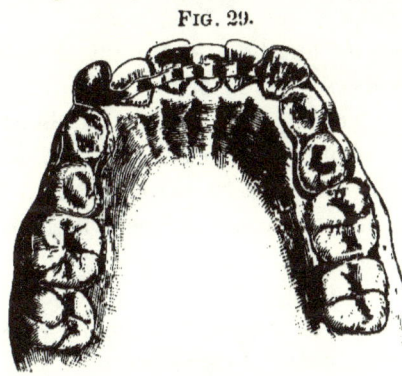

Fig. 29.

Corrugated Band Regulator (Byrnes).

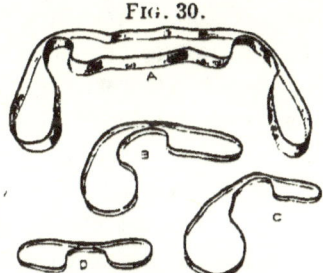

Fig. 30.

Band Regulators.

Fig. 29 shows another form of corrugated band used to press forward the inferior incisors. Two teeth on each side are here used as anchorages, being connected by a single continuous band.

Fig. 30 illustrates other forms of the band regulator made from extremely narrow strips of gold.

THE MAGILL BAND.

This device, while not properly constituting a method, is considered here, because through its great value it has come to be an important factor in several methods of regulating devised since its introduction. Dr. W. E. Magill, having in common with other practitioners experienced the difficulty of attaching regulating appliances to the natural teeth in such a way that they would have a firm hold and not slip, devised the following plan of meeting and overcoming the difficulty :—

From a piece of platinum, German silver or platinized silver plate, No. 28, (B. and S.) gauge in thickness, he cut a strip about a line in width and bending it to conform to the shape of the tooth soldered it at the point where the ends overlapped, thus converting it into a band or ferrule. After attaching to this band any studs, pins or hooks that the case demanded, it was lined with oxy-chloride of zinc and slipped over the dried tooth to a point about midway between the cutting edge and neck.

Since the introduction of phosphate of zinc, it has been found to be a far better medium for the attachment of the band to the tooth than the oxy-chloride of zinc, formerly used. Once in position, the cement will harden in about five minutes, after which no ordinary force will be able to dislodge it. If a wire spring is intended to rest against and press upon a banded tooth, a hole or pit should be drilled in the band at a suitable point, before it is cemented in place. If rubber bands or ligatures are to be employed, suitable provision for their easy attachment may be made by previously soldering to the band a small gold hook, or a headed platinum pin taken from a vulcanite tooth. Where a jackscrew is to be used in the moving of a tooth, an abutment of platinum should be soldered to the band encircling the resisting tooth, and then be slotted to receive one end of the screw. The band of the tooth to be moved should also be reënforced and drilled to accommodate the point of the screw.

When the operation is completed, or when for any cause it may be desired to remove the band, it is easily accomplished by protecting the enamel at the cutting edge of the tooth with a folded napkin or piece of chamois skin, and placing one beak of a pair of pliers upon it and the other upon the upper edge of the band, the closure of the hand will dislodge the appliance without in the least marring or altering its form. By this simple invention, one of the greatest difficulties hitherto experienced in regulating has been overcome, and its devising has almost introduced a new era in regulating. For the purpose intended, there is nothing that approaches it in efficiency.

Before its introduction, attachment to the tooth to be moved was usually effected by means of a ligature ingeniously applied and made fast by some form of knot, or a pit or hole was drilled into the substance of the tooth to receive the point of a screw or other device and prevent it from slipping. The knots would often slip, and the drilling of pits was objectionable, so that the difficulties of securement were not overcome until the invention of this band.

By its use absolutely secure attachment and anchorage are obtained, and the moving of teeth is accomplished with far greater exactness than had previously been possible. When attachment was made by ligature, it was often necessary that the ligature should encircle the tooth at its neck, and when not necessary to place it there it would often slip into that position owing to the shape of the tooth. The irritation of the soft tissues thus produced was frequently the cause of much pain to the patient. The Magill band obviates this by preventing any fixtures attached to it from coming in contact with the delicate and sensitive mucous membrane of the gum.

Indeed, the author has found that by its use nearly all the pain of regulating has been done away with, for the pain attendant upon regulating by the old methods was caused not so much by the slight irritation induced by the moving tooth, as by the impingement of ligatures, rubber bands and other appliances upon the soft tissues. The Magill band may therefore, we think, be credited with having done more to modify the pain accompanying regulation than any other device ever introduced.

In some methods of regulating, such as Farrar's and Patrick's, attachment is made to the teeth by means of an open band of gold secured to the teeth by a nut and bolt operating upon the free ends of the band. Such device, while valuable, is more complicated, cumbersome and less cleanly than the Magill band. It is also open to the objection previously noted, that of allowing the secretions to remain between the tooth and band.

Several of the author's methods of modifying the form of the band by means of attachments to increase its usefulness, are illustrated in Part III.

ANGLE'S METHOD.

This method of regulating was first brought to the notice of the profession by its originator, Prof. Edward H. Angle, in a paper read before the dental section of the Ninth International Medical Congress, held in Washington, D. C., September, 1887.

The appliances used in this method are mostly made from German silver, although the levers are of steel and the retaining wire of gold. German silver is strong, easily adapted and inexpensive, while the steel piano-wire combines strength and elasticity with lightness and delicacy.

Power is obtained by the well-known mechanical principles of the screw and lever, while support or resistance is gained by firmly attaching the parts to the teeth by the Magill band, which is cemented in place, or by an adjustable clamp band.

The appliances are few in number, simple in design, and easily applied; qualities that add materially to the value of any device for general use. Prof. Angle, in describing his method, says:—

"Fig. 31 shows the simple appliances from which all the various combinations used in the original method may be made. "A" is a large traction screw encased in its accompanying tube, and used for pulling where the resistance is great. "B" is a smaller traction screw, used in the same way where the resistance is slight, or where from any reason a delicate appliance is desired. "C" and "D" are tubes which are soldered to bands placed upon the teeth to be moved, into which the ends of the traction screws are hooked. "J" is a jack-screw, used for pushing, the end of which is beaten flat. "E" is an extra piece of tubing, by means of which a longer jack-screw can be made. "F" and "H" are coils of band material of different thicknesses. "G" is a gold wire used in retaining the teeth and also to assist

MATERIALS AND METHODS.   101

in securing anchorage in some cases, and "RR" are small retaining tubes, into which the retaining wire accurately fits,

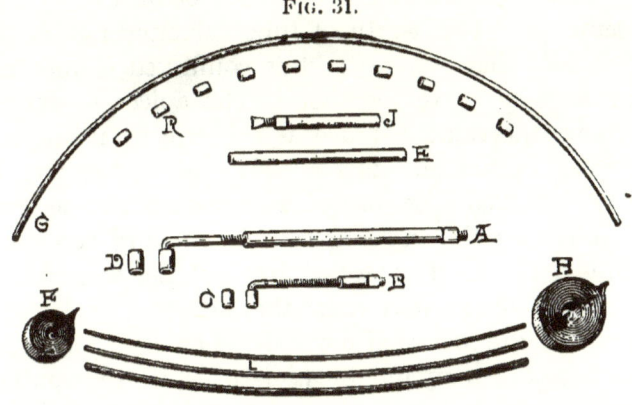

Fig. 31.

Angle's Appliances.

and are designed to be soldered to bands. "LL" are piano-wire levers of varying sizes, giving different degrees of power.

"Aside from the advantages of simplicity, efficiency and cleanliness, which are insured by these appliances, a still greater desideratum is gained by means of the mechanical principles observed in their construction. Stationary anchorage and non-relinquishment of pressure are prominent features of this method, and are certainly secured almost to perfection.

"The means by which one or more teeth are held perfectly stationary, while serving as an anchorage or base of resistance for the application of force is quite simple and peculiar to this method.

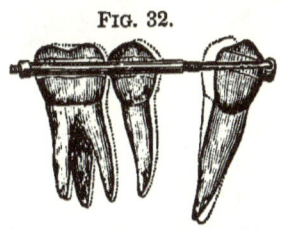

Fig. 32.

Stationary Anchorage. (Angle)

"One or more teeth are banded, as shown in Fig. 32. Soldered to the bands is a tube of some length. Through this tube a rigid shaft, threaded at one end and bent to a right angle at the other, is passed to a tooth to be moved. On turning the nut the natural tendency would be to tip the anchor teeth forward in their sockets; but they cannot tip thus, because of their rigid connection, and the length of tube surrounding the shaft. It is evident that two teeth thus connected cannot move, except as they move together. The apices of the roots must move the same distance as the crowns, if any movement at all occurs, and this is well nigh impossible. The tooth to be moved is connected with the shaft in such manner that it may tip, and responds by moving according to the force applied. The dotted lines of the diagram show the direction of any movement that occurs.

"A few of the principal movements are selected for illustration from the many modifications of which the appliances are capable.

"The application and operation of the direct screw is shown in Fig. 33. A firm anchorage for the resistance of the screw is obtained by banding and tubing the left cuspid, and passing through the tube a piece of gold or German silver wire long enough to extend to and rest against adjoining teeth. The opposite cuspid is banded, and a retaining tube soldered to the labial

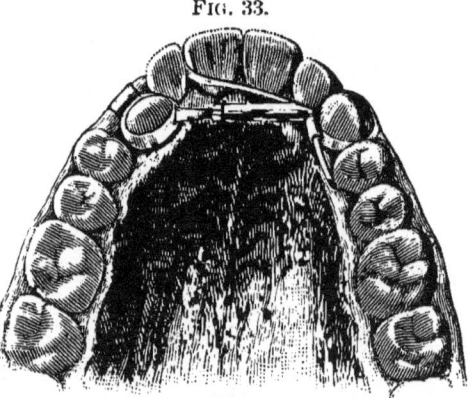

Fig. 33.

Re-enforced Anchorage. (Angle.)

surface. The lingual surface has a slot cut in it to receive the flat end of the jack-screw. The other end of the tube, in which the screw plays, is so filed that it rests securely against the reënforcement wire and the tube upon the lingual surface of the cuspid band. After being brought into position the tooth is held in place by passing a short piece of gold wire through the retaining tube on the labial surface, which is left in place until the tooth is firmly set in its new position.

"The backward movement of teeth in the line of the arch is accomplished by the appliance shown in Fig. 34. The second bicuspid and first molar are banded, and the tube of the heavy traction screw rigidly soldered to the bands. The cuspid to be moved is banded, and a short section of tubing soldered to it to receive the end of the traction screw.

On turning the nut traction is produced and the cuspid drawn into place. The cuspid is kept from being rotated while it is being moved backward, by means of the short tube accurately fitting the right angled end of the traction screw.

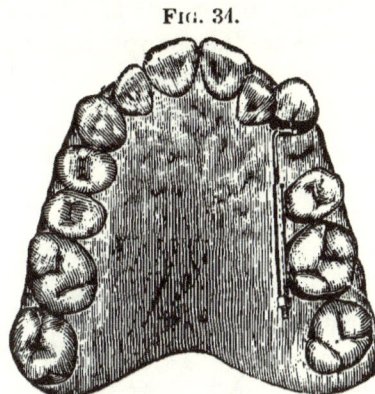

Fig. 34.

Retraction of Cuspid. (Angle.)

"Another outward movement of a tooth by means of the jack-screw is shown in Fig. 35. The second bicuspid is made the principal anchorage, against which the base of the tube rests. The band encircling the lateral incisor has a slot cut in it to receive the end of the jackscrew. The anchorage is reënforced by means of a wire loop, which hooks into tubes upon the adjoining central and cuspid, and is looped over a spur upon the body of the jack-screw tube.

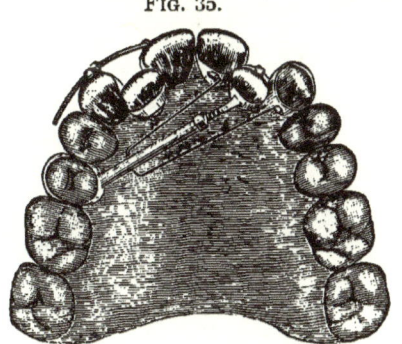

Fig. 35.

Reciprocal Anchorage. (Angle.)

The central and cuspid cannot be pushed outward on account of this reënforcement, and three teeth constitute the anchorage instead of one. The several parts of this appliance are shown in Fig. 36.

FIG. 36.

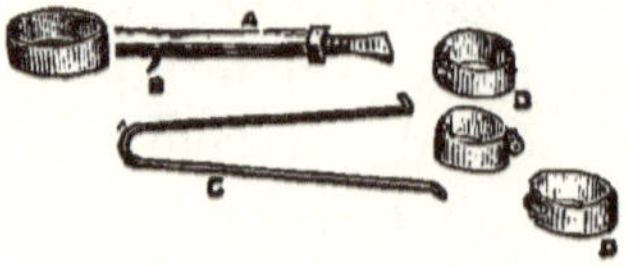

"Outward movement, as accomplished by another simple means, is as follows: A thin strip of band material is looped about the malposed tooth, the ends resting upon the labial surfaces of the adjoining teeth. To one end of this strip is soldered a tube placed vertically, while to the other end a similar tube is attached horizontally. Into these tubes the small traction screw is placed, being bent to conform to the shape of the arch, and used in this case to push instead of pull. The parts of this device are shown separately in Fig. 37. The manner of retaining the teeth in position, after correction, is shown in Fig. 38.

FIG. 37.

Device for Lateral Movement.

FIG. 38.

Retention.

"Rotation by this method, as in most others, is accomplished by the leverage and elasticity of a metallic bar or wire attached to the tooth to be rotated, and then sprung around to some firmer tooth or teeth at a distance. Fig. 39 shows a lateral to be rotated, and the appliance in position by which it may be accomplished. The lateral is banded and tubed as shown in the cut. The second bicuspid is also banded, and to secure greater resistance, the two adjoining teeth are made to assist by means of a wire which passes through a tube on the palatine surface and rests against the first bicuspid and first molar. On the buccal side of this same band, the ends of the band material

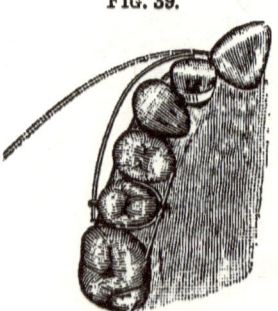

FIG. 39.

Rotation. (Angle.)

are shaped into a latch or hook, with which the rotating lever engages when it is sprung around. The several parts of this appliance are shown in Fig. 40. After the tooth is in position, it is retained by means of a short wire passing through the tube, and extending upon the central, as seen in Fig. 41. This wire is kept in place by a small pin, which is tightly fitted in a small hole drilled through both tube and wire, as shown.

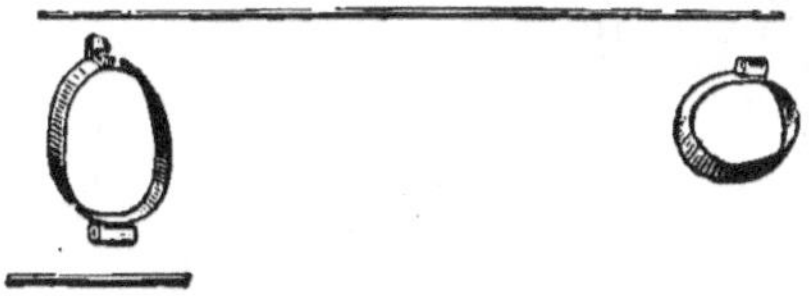

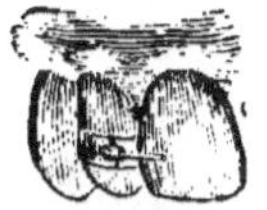

FIG. 40.

FIG. 41.

Retainer.

"When two teeth are to be rotated in opposite directions at the same time, as the central incisors, double rotation may be accomplished by one appliance, as shown in Fig. 42. Both teeth are banded, and a tube soldered to each band, one being horizontal and the other vertical. A piece of piano-wire is bent to a right angle at one end, and then placed in position as seen in Fig. 43. The tendency of the wire to straighten itself, will rotate both teeth at once. When in position they are retained by substituting a non-elastic gold wire for the piano-wire.

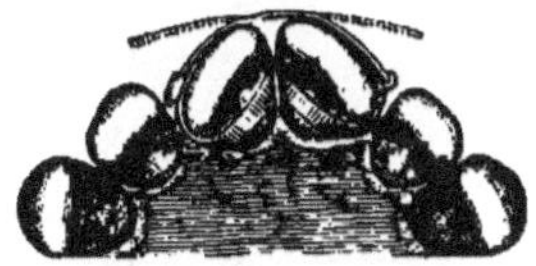

FIG. 42.

Double Rotation. (Angle.)

"Expansion of the arch is accomplished by banding and tubing the first and last teeth of those to be moved, on each side, and connecting them by means of a wire passed through the tubes. To these wires, at suitable and varying distances, are soldered short tubes to accommodate the ends of the piano-wire spring which is bent to conform somewhat to the shape of the arch. While the spring does not give us the power and direct action of the jack-screw, it is in many cases sufficient

FIG. 43.

Double Rotation. (Angle.)

and avoids interference with the tongue which necessarily accompanies the use of the latter.

"Fig. 44 shows the appliance in position, which is as applicable to the lower teeth as the upper."

Retention is anticipated and provided for, by means of the tubed band, while the pin device for locking lever and tube together, is both novel and ingenious. Aside from these, the method contains so many ingenious modifications of previously known devices (as the screw and band), and is composed of parts so simple and direct in their action, that it must necessarily commend itself to all engaged in this line of practice. Other illustrations of this method are shown throughout Part III.

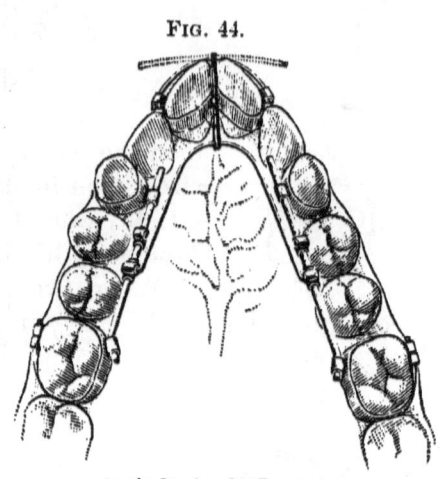

FIG. 44.

Angle Device for Expansion.

The various parts of the Angle appliances may be obtained from the inventor, or from the dental supply houses.*

COFFIN'S METHOD.

In a paper read before the Dental Section of the International Medical Congress, held in London, in August, 1881, Mr. Walter H. Coffin explained his method of correcting irregularity of the teeth. The method was devised by his father, and had been in use by father and son for twenty-five years. It was termed the "Expansion Method," because in nearly all cases coming under their care a certain amount of expansion had been found necessary in connection with other desired movements.

---

* Directions for constructing the Angle and other metallic appliances will be found in Part IV, Chap. V.

The construction of the appliance and the principle upon which it acts are exceedingly simple. The power is derived from the elasticity of piano-forte wire, attached in various ways to a vulcanite plate which covers the arch (in an upper case) and envelopes the posterior teeth on either side to give it firmness and fixedness in position. When it is desired to expand the superior arch, the wire is bent into the following form ᘜᘞ, lying on top of the plate with the ends embedded in it.

To produce lateral expansion in the lower jaw, the form of the appliance is necessarily different. A simple vulcanite plate is made in horse-shoe form, fitting the gum and lingual surfaces of the teeth, and capping the molars and bicuspids. On the lingual surface of this plate, lie two pieces of piano-wire suitably curved, with their ends embedded in the rubber.

Each of these plates, when completed, is sawn in two along the median line, thus allowing the tension of the wire to be increased from time to time by spreading apart the sections of the plate.

The piano-forte wire used may be obtained from piano factories or from dealers in dental supplies. It is simply wire made from the best quality of steel, drawn to size through draw-plates. The quality of the steel, as well as the toughness of the wire, is greatly improved by the successive drawings to which it has been subjected. For ordinary cases Mr. Coffin recommends that the diameter of the wire be between three and four one-hundredths of an inch. A lighter or heavier number will yield respectively less or greater pressure.

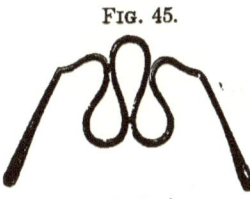

Fig. 45. Coffin Spring.

In use it should not be annealed, but bent to shape as it comes. Mr. Coffin recommends that the wire be tinned after being bent to shape, to prevent oxidation in the mouth, but this does not appear to be necessary.

A wire suitably bent to produce expansion of the superior arch is represented by Fig. 45.

The details of the construction of an expansion plate for the superior jaw, are as follows: From an accurate impression of the jaw and teeth, taken with plaster or modelling compound, a plaster model is obtained. Upon this a wax base-plate is fashioned, to cover all parts intended to be covered by the completed plate. The suitably bent wire is now further shaped so that it will lie upon the exposed surface of the base-plate and conform to it as closely as possible in outline. After the ends of the wire are attached to the base-plate by means of additional wax, a piece of tin-foil (No. 60) is slipped between the wire and the plate and its corners bent, so that the plaster when poured into the flask will grasp and remove it with the wire. The foil is placed there so that the plate will have a polished surface under the wire after vulcanization. The wax base-plate should now be smoothed with a spatula and flasked in the usual manner. In separating the flask, the wire and tin-foil will come away with the upper half, while the model will remain in the lower. After removing the wax and packing the rubber, the case is vulcanized, after which it is polished. The completed piece should now be properly fitted to the patient's mouth, and the rubber covering the masticating surfaces of the posterior teeth so filed and dressed that the cusps of the occluding teeth will all strike the rubber at the same time.

However many or few of the natural teeth be covered the last ones in the arch must always be included, as otherwise they will elongate through non-occlusion and thus seriously impair the usefulness of the masticatory apparatus. After the plate has been fitted it should be sawn in two with a jeweler's fine saw, the edges made smooth and slightly rounded, and the case introduced into the mouth.

It is desirable to have the patient wear the plate for a day without enlargement, after which, at intervals of a day or two, the tension of the wires should be increased by pulling the halves of the plate apart sufficiently to slightly increase the space between them. When the wire is heavy, as is necessary where great force is to be exerted, it can be best formed into shape and afterwards altered as required by means of the ordinary clasp-bending pliers. The construction of the lower plate is substantially the same, but the wires lie against the plate in a continuous smooth curve, instead of being corrugated.

Figs. 46 and 47 represent an upper and lower expansion plate as described. For cases where expansion is not needed, but simply the moving of one or more teeth, Mr. Coffin uses a solid rubber plate with wires so placed as to produce the desired movements. The construction of this form of plate is the same as those just described, with the exception of the shape and arrangement of the wires and the non-separation of the plate.

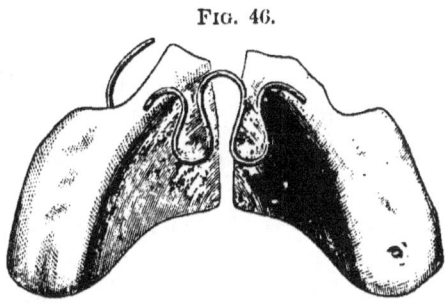

FIG. 46.

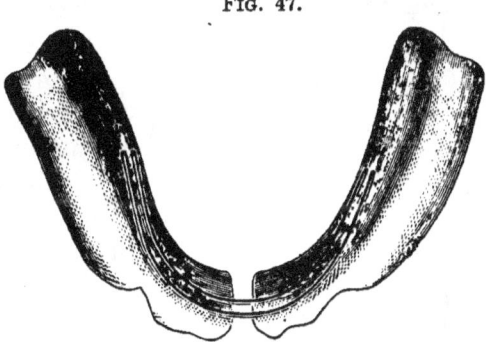

FIG. 47.

Coffin's Expansion Plates.

A single long piece of wire, bent at right angles near one end and flattened at the other, is embedded at its flattened end into the plate, while the other end, and a long portion besides is free and lies in close contact with the plate. Before the wire is attached to the wax base-plate, the plaster tooth representing the one to be moved should be cut away close to its neck and the bent end of the wire laid upon it so as to cover the entire diameter of the stub tooth. In this position it is vulcanized to the plate.

When the plate is introduced, the wire will have to be drawn back with an instrument or string before the plate will go into position. Once in place and the wire released continuous pressure will be exerted on the malposed tooth. After the tension of the wire has been lessened by the moving of the tooth, it may be increased either by bending the wire where it enters the plate or by cutting it out and re-setting in a different position.

Another and very convenient way of lengthening the wires to follow the moving tooth, is to slip a section of platinum or German silver tubing over the end of the wire and soft-solder it in position.

Where a tooth is to be pressed outward the wire is anchored in the palatal portion of the plate, but where a tooth is to be moved from without inward, the wire should be attached to that portion of the plate covering the buccal surfaces of the molars.

Rotation is accomplished by combining the two movements; that is, by having one wire on the palatine surface to press against one angle of the tooth, and another on the buccal surface to press against the opposite angle.

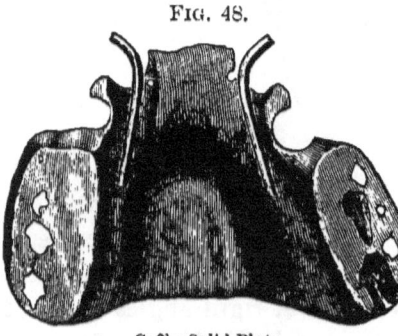

Fig. 48.

Coffin Solid Plate.

Two wires can be inserted to operate on two teeth at the same time, either in similar or opposite directions. Fig. 48 represents a plate made to press outward two lateral incisors.

Many modifications of the Coffin plate have been devised by different practitioners, some of which are shown in Part III.

The originator claims for his method and appliance, simplicity, ease of construction and inexpensiveness, almost universal range of application, perfect control of force applied and direct action, comparative painlessness from non-irritation of the soft tissues, perfect fixedness and least unsightliness, ease of removal for cleansing, and little interference with speech and mastication.

Dr. E. S. Talbot has designed a modification of the Coffin piano-wire spring, which consists in converting it into a coil at some point of its length, thus adding, it is claimed, greater elasticity and a wider range of applicability. Unlike the Coffin spring it may be used without a rubber plate and without being permanently attached to any appliance. The coil is formed by bending the wire around a mandril firmly driven into the bench or properly secured in a vise.

The arms may be bent or cut to any length to suit the case in hand. They may be used in connection with a rubber plate, or with bands of gold or platinum fastened to the teeth with zinc-phosphate. With holes properly drilled in the bands or plate and the arms fitted into them, the spring will stay in position. When the spring is used without a plate it may be well to fasten the wire to some of the teeth to prevent its being swallowed. Fig. 49 illustrates the coil spring in some of its forms. To prevent the spring from rocking in the mouth the coil is usually made to press over a button or post suitably placed on the plate for that purpose.

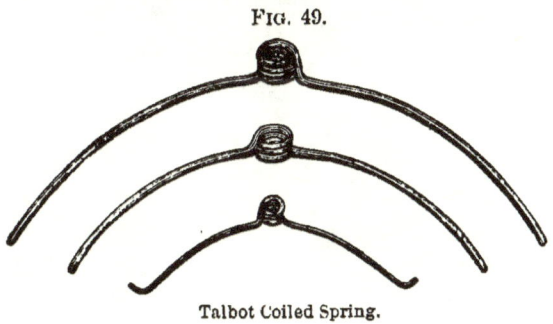

FIG. 49.

Talbot Coiled Spring.

The coiled spring, in many cases, possesses advantages over

the plain spring because it can be effectively used where the other cannot. It is also more easily regulated as to tension, and can be readily replaced by a weaker or stronger one should the case require it.

JACKSON'S METHOD.

Appreciating the value of piano-wire as a power-yielding material, as shown in the Coffin method, and realizing the advantage in most cases of dispensing with the use of a plate, Dr. V. H. Jackson was led to devise a method of constructing regulating appliances in which piano-wire was the principal and almost the only material employed.

By suitably bending a length of this wire, of medium thickness, in such a way as to pass around the buccal and lingual surfaces of all the teeth in one of the arches and joining these portions at convenient distances by short connecting wires, a "crib" or skeleton-wire fixture was formed that hugged the teeth and held itself firmly in place.

To this, as a foundation, additional wires were attached of such length and shape as to bear and produce pressure upon any teeth in the same arch as it was desired to bring into proper position.

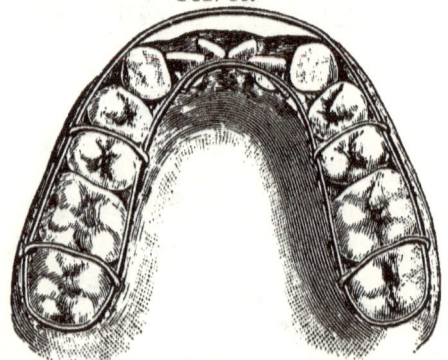

Fig. 50.

Crib. (Jackson.)

Fig. 50 shows the general appearance of the "crib" in its simplest form.

In constructing the appliance, the plaster teeth of the model are first scraped near their necks on both the buccal and lingual surfaces so that the crib, when formed, will have to be sprung into place. The wire is now bent by means of flat- and round-nosed pliers so as to conform to the outline of the teeth and touch all of the included ones at their necks.

To keep the crib from impinging upon and irritating the gum, short wires (as before stated) are formed to lie in the depressions between the masticating surfaces of certain teeth and are attached to the main wire upon both the buccal and lingual sides. These connecting wires are joined to the base wire by having their ends bent so as to grasp them, after which the joints are secured by means of soft-solder fused by either the blow-pipe or soldering iron while the parts are in position on the model. Before soldering, the parts will have to be touched with dilute muriate of zinc, commonly known as soldering fluid. Wrapping the joint with fine copper wire before soldering greatly facilitates the operation.

The crib once properly formed, additional wires for producing pressure at any point and in any desired direction are added to it in the same manner.

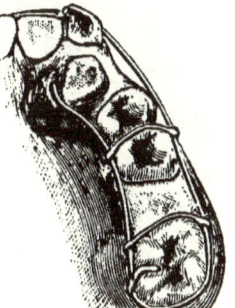

FIG. 51.

Side Crib. (Jackson.)

Fig. 51 shows a crib formed for and attached to but one side of the arch for the purpose of forcing a cuspid outward and a lateral inward into line at the same time.

In some cases the end of the wire producing pressure is best secured in position by being soldered to a band to be cemented to the tooth to be moved, as shown in Fig. 52.

While the appliance thus constructed is firmly held in place by hugging the teeth above their most prominent portions it is at the same time readily removed for the purpose of bending the wire springs or for alterations or new attachments.

Dr. Jackson has recently simplified and improved his appliances by discarding the crib formed of a continuous piece of wire (which was oftentimes difficult to construct) and obtaining his anchorage by wire and metal attachments to individual teeth instead, as shown in Fig. 53.

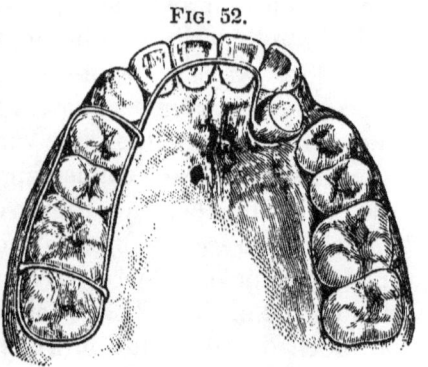

Fig. 52.

Crib and Band. (Jackson.)

In constructing these anchorage appliances, he first cuts from thin gold, block tin, tinned copper, German silver or Tagger's tin a piece large enough to cover the lingual portion of the anchor tooth and contours it with the contouring pliers used in crown- and bridge-work. A wire crib for the same tooth is then made from a piece of No. 20 piano-wire by "first bending it at right angles (Fig. 54), leaving the width between the parallel sides equal to the antero-posterior width of the tooth to be clasped. The part that is to clasp the neck of the tooth is then so bent with clasp-benders that it will be perfectly adapted to the curve of the labial side of the tooth. (Fig. 55.) Both arms of the wire are then bent at nearly a right angle at a proper distance to cause them to pass over the grinding surface of the tooth, and again bent in the same manner to extend toward the neck of the tooth on the lingual side. (Fig. 56.)

Fig. 53.

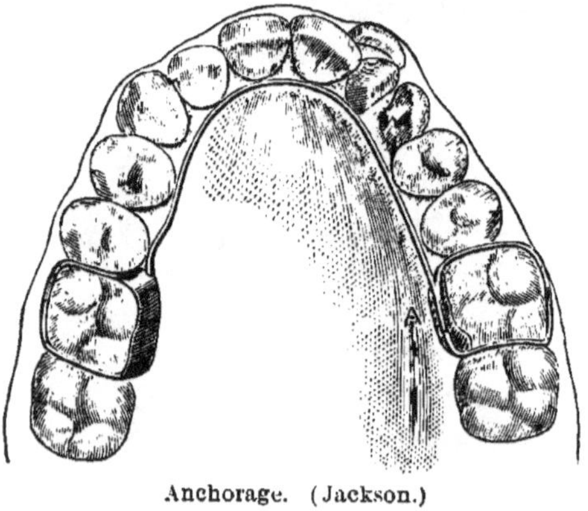

Anchorage. (Jackson.)

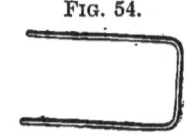

Fig. 54.

"The ends are next bent toward each other near the gum line over the piece of metal previously described, as seen at A in Fig. 53, and tacked with soft solder."

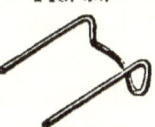

Fig. 55.

If the wire spring is to be attached to the teeth on the opposite side of the arch a similarly constructed crib should be made for that side. With these two cribs in position on the plaster model the connecting wire, after being suitably shaped, is laid in position and firmly held while all are joined together with solder.

Fig. 56.

The soldering is most conveniently accomplished by moistening the parts with dilute muriate of zinc, laying upon each joint a piece of soft solder of suitable size and fusing with a soldering iron. After this any wire springs that may be needed are attached in the same manner.

The entire appliance being thus formed of separate parts and joined together while in position on the model assures accuracy of fit that could not well be obtained in any other manner.

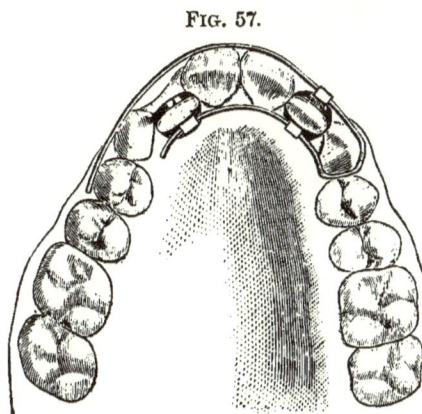

Fig. 57.

Wire and Band Appliance. (Jackson.)

In other cases, where it seems advisable, instead of a crib made of plate and wire as described, attachment to the anchor teeth is made by means of a metal band or collar encircling the tooth, to which sections of metal tubing or lugs are attached for the accommodation of the spring wire, as shown in Fig. 57.

Some of the numerous ways in which these combination appliances may be adapted to the correction of many forms of irregularity are shown in connection with the practical treatment of cases in Part III.

Dr. Jackson claims for his method that the appliances are easily retained, cause no inconvenience and do not interfere with articulation, even when used in both the upper and lower arches at the same time. In addition, the model is not injured in making the appliance and so may be preserved for future measurements and study.

PART III.

SPECIFIC FORMS OF IRREGULARITY AND THEIR TREATMENT.

While principles and methods may be well understood, illustrations of their application in certain forms of irregularity will be necessary in order that the student may properly comprehend their practical relationship.

So far as ease or difficulty of treatment is concerned, cases of irregularity are naturally divided into two general classes; in one the cases are brought to our notice as soon as the irregularity begins to manifest itself, while in the other the deformity is fully establishsd and confirmed before presentation for treatment. In the first class, occurring usually in children, we have the advantages of easy movement and freedom from complications; while in the second, we have to contend with slow and difficult movement and a variety of unfavoring conditions.

For these reasons it is deemed advisable to treat of certain forms of irregularity, especially those involving the six anterior teeth of each jaw, under separate heads, according as they present before or after dentition is complete, for the treatment in one case will vary considerably from that required in the other.

CHAPTER I.

INCISOR TEEEH ERUPTING OUTSIDE OR INSIDE OF THE ARCH.

Reference has already been made to the fact that normally the permanent inferior incisors erupt inside of the arch and posteriorly to the deciduous ones, while the permanent superior incisors erupt outside of their deciduous predecessors. From the limited space allotted to them, there is a stronger tendency to irregularity on the part of the lower incisors than there is on the part of the more favorably located superior ones, although the latter are also often found in a crowded condition, sometimes complicated with torsion.

So long as the inferior ones are within the arch, even though irregularly arranged, they will usually need no attention on our part until dentition is complete, and when that time arrives it will generally be found that nature has almost, if not entirely, corrected the condition.

So, also, where some of the superior incisors erupt slightly outside of the arch, they being still in line with spaces between them, we need not interfere, for in most cases the force exerted by the lips and the erupting cuspids will bring them into normal position and relationship.

It not unfrequently happens, however, that from some cause a superior incisor is deflected in its eruption and appears inside of the arch, or that an inferior incisor is found to erupt outside of the arch. In either case, treatment is indicated as soon as the irregular tooth or teeth are sufficiently erupted to enable us to bring the proper force to bear upon them.

One of the earliest methods employed for releasing an inlocked superior incisor was by the use of what was known as the "saddle and inclined plane," one form of which is shown in Fig. 58.

The *saddle* was usually formed of metal, struck up to fit and cover all of the lower incisor teeth. To this, at some point of the ridge, was soldered an inclined piece of heavy metal so arranged that the inlocked tooth would strike upon it in mastication and be forced outward into line.

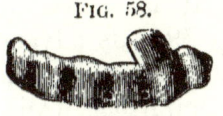

Fig. 58. Inclined Plane.

Later the appliance was often made of vulcanite, and while in either form it generally answered the purpose of correcting the simple irregularity, it was objectionable on account of its size and because it was removable and thus liable to be lost or laid aside and not worn.

A modification of and improvement upon the old form, retaining its virtues and obviating its disadvantages, was devised by the author many years ago. By its use, when attached to a single tooth, a double movement is produced, for while by the action of the plane the superior inlocked tooth is moved outward, the lower outstanding one, to which the plane is attached, is moved inward. When it is not desired to move the lower tooth it can be prevented by making the appliance to include two or more teeth and thus offer more resistance.

It is constructed as follows: A band of thin platinum, gold or German silver plate (No. 29, B. and S. gauge) is bent to encircle and fit the protruding lower incisor, and the ends soldered. A piece of ordinary gold plate is then bent double to form a plane, and spread apart at its ends to grasp the band on the lingual and labial surfaces, to which it is soldered. It is next placed upon the tooth to see that the adjustment is correct, removed, lined with phosphate of zinc, and pressed permanently into position. If the teeth are in close contact it is well to allow the fixture to be worn a day previous to cementing, for then the teeth will have been pressed apart and the replacement with cement will be more easily accomplished. The cement not only lines the band, but fills all the space between the plane and the tooth, thus giving greater resistance and strength in biting. It is shown in position and separately, in Fig. 59. Its advantages are its small size and absolute fixedness. When the correction has been accomplished, it will be necessary to cut the band in order to remove it. Two objections have been urged against the employment of inclined planes in any form: one, that by thus opening the bite, the posterior teeth will elongate; the other, that the patient may avoid biting upon the plane and thus defeat our object. These objections have no real validity, as is shown by actual experience.

Fig. 59. Fixed Plane.

The short time that the bite is open, usually only a week or two, is not long enough to permit of any perceptible elongation, while the patient must and does bite upon the plane in mastication, because it is the only point where occlusion is possible.

Another plan of accomplishing the same end has been suggested by Prof. C. N. Peirce. He attaches ligatures to several or all of the lower incisors, and makes these fast to the molars on either side. The ligatures being attached and drawn tight while dry, will, under moisture, contract and draw the incisors inward. This operation is continued until the lower incisors reach a position inside or back of the malposed superior ones. The ligatures are then removed, and the lower teeth, in gradually resuming the position they formerly occupied, will carry the inlocked superior ones with them.

Where, for any cause, it is desirable to confine the means of correction to the jaw in which the irregularity exists, as, for instance, where the superior laterals are inlocked, a simple plan is to take a piece of platinized gold, about one-eighth of an inch in width and long enough to more than cover the four incisors, and punch or drill four holes in it, two opposite each of the laterals. The bar being laid in position on the labial surfaces of the centrals, the laterals are securely ligated to it, the thread passing through the holes. The spring of the bar and the contraction of the moist ligatures, will move the laterals into position in a short time, the ligatures being renewed every two or three days.

A more satisfactory way of performing this operation is to solder one end of the bar to a platinum band made to encircle one of the laterals and attached to it by zinc cement. Arranged in this way, the bar has but one free end, which is more readily ligated to the other lateral.

Fig. 60 illustrates an appliance of this character, that was used to bring out into position two superior laterals in the mouth of a girl ten years of age. The case was complicated by one of the centrals being slightly turned upon its axis.

A platinum band or collar was made to fit the right lateral, and to its labial surface was soldered one end of a bar of spring gold, long enough to extend over the centrals and cover the opposite lateral. The bar was converted into a hook at its free end and so shaped that in its course it touched only the prominent edge of the twisted central. The band was then cemented to the right lateral, and a section of small rubber tubing passed under the left lateral and caught in the hook. The appliance thus operated in two ways: First, to bring the laterals out into line; and next, to press backward and inward the protruding corner of the central.

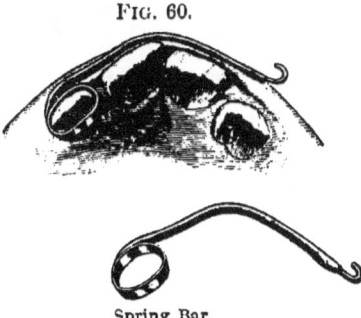

FIG. 60.

Spring Bar.

Another way of securing the same result is by the use of a Coffin plate and suitably-shaped extension wires, as shown in Fig. 61.

The rubber plate is made to cover the arch and enclose several bicuspids or molars on each side. In each of the buccal portions of the plate a piece of piano-wire is imbedded, which extends forward clear of the teeth and terminates in a curve or hook opposite the tooth to be moved outward. A section of rubber tubing is slipped over the tooth and caught upon the hook. The elasticity of the rubber added to the spring of the metal will rapidly draw the tooth outward provided there is sufficient space in the arch to accommodate it.

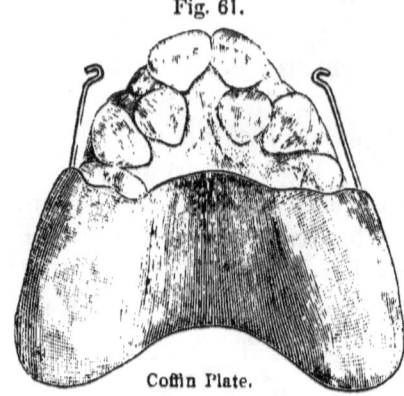
Fig. 61.
Coffin Plate.

As already stated, slight spaces existing between the superior incisors when recently erupted need give us no concern provided they are in the normal line of the arch; but it often happens that in addition to the spacing one or more of them is, to a greater or less extent, turned upon its axis, as shown in Fig. 62.

Fig. 62.
Torsion with Space.

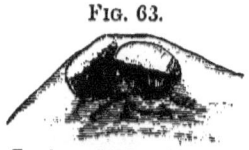
Fig. 63.
Torsion with Overlapping.

In other cases the teeth may be in contact, while one of them is twisted and overlapping its neighbor, as shown in Fig. 63. In either case it is quite probable that the cutting edge of the turned tooth will occlude with the corresponding surface of the one in the opposite jaw at an angle, and thus either prevent full eruption of one or the other of the teeth, or temporarily open the bite and favor undue elongation of posterior teeth.

Both of these forms of irregularity should receive immediate attention, for at an early age correction is easily accomplished. Were the condition to remain unchanged, it would necessarily become more complicated from partial closure of the space caused by the lateral pressure that would be exerted during the eruption of neighboring teeth.

Rotation of these teeth, as well as of others, may be accomplished by one of the many methods described in Chapter VI.

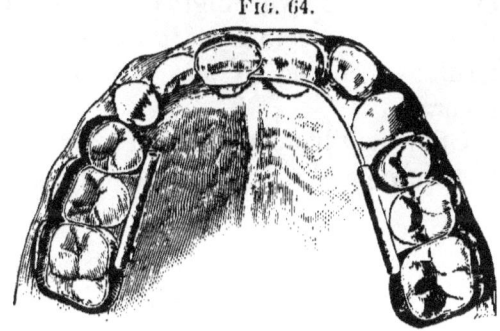

Fig. 64.

Tube, Band and Spring Appliance. (Matteson).

Dr. Matteson accomplishes the same result without the employment of a rubber plate. He prefers to band the first deciduous and first permanent molars and joining these bands by a connecting strip on the buccal surface, and a piece of round tubing closed at one end on the palatal surface, as shown in Fig. 64.

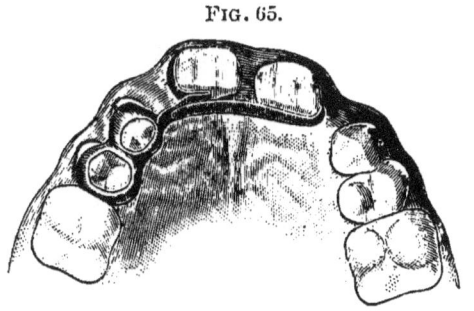

Fig. 65.

Simple Spring. (Matteson).

The incisor to be moved forward has a band of gold or platinum cemented to it, and to this band, on the palatal surface, is soldered a U-shaped lug.

By inserting a piece of thin piano-wire into the tube and springing its outer end into the lug on the incisor band the tooth is readily forced into position.

Instead of the tube and wire he sometimes employs a thin strip of platinized gold soldered to the bands and made to rest and press against the in-lying tooth, as show in Fig. 65.

The single appliance may be used to press forward both of the incisors by arranging the strip of spring gold to press upon but one tooth until it is in place, and then altering its form by bending so that it will exert its force upon the other.

Other appliances of somewhat similar character will readily suggest themselves to an inventive mind.

CHAPTER II.

DELAYED OR MAL-ERUPTION OF THE PERMANENT CUSPIDS.

The third molars excepted, the superior cuspids are usually the last teeth of the permanent set to erupt, and they almost invariably make their appearance outside of the arch. When there is room in the arch for their accommodation and they erupt directly outside of it, we may feel assured that in due time they will find their way into place unaided. Where, however, they erupt over the lateral incisors, as is sometimes the case, and these incisors are in consequence being forced inward, it becomes necessary for us to interfere and endeavor to draw the cuspids toward their proper places. This is usually not a difficult matter when the cuspid crown is far enough erupted to enable us to exert pressure upon it. In such a case, by cementing a Magill band to the cuspid and another to the second bicuspid or first molar, each having a pin or hook attached to its buccal surface, a rubber ring extending from hook to hook will in a short time draw the cuspid back to a position opposite the space it is to occupy, as illustrated in Fig. 66.

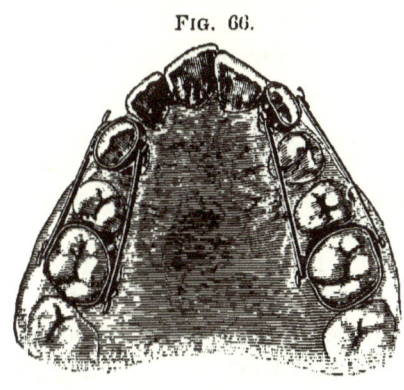

Fig. 66.

It sometimes happens, however, that the cuspids are tardy in their eruption and fail to assume their positions in the arch at the time they are needed to complete the row and prevent the incisors and bicuspids from encroaching upon the space the cuspids are to occupy. In such cases it is generally advisable to hasten their eruption by the application of tractile force in some manner. Where one-half of the crown is through the gum we can attach to it a Magill band with a pin, hook or other projection upon it, and by its assistance readily apply power to the tooth.

The author has had several cases where elongation of the cuspid was called for, when only the cusp of the tooth was visible through the gum. Here the application of a cemented band was out of the question, and attachment to the tooth had to be gained in another way. The difficulty was solved by tying a silk ligature in a half knot, passing it over the projecting cusp, and then with a small, flat plugger, forcing this ligature up under both gum and alveolus until it encircled the neck of the tooth, when it was drawn tight and made fast with a surgeon's knot. A very small gold ring, with a centre only large enough to admit of the passage of silk floss, was then slipped over one of the ends of the ligature and tied so that it would lie upon the labial face of the tooth near the gum. This ring was allowed to remain without change until the tooth was drawn into position. A delicate vulcanite plate was constructed to fit the arch, and extend into the space between the lateral and first bicuspid. At this latter point the plate was thickened until it was nearly on a level with the cutting edges of the adjoining teeth, and made concave on its most prominent part. A rubber spur was also formed on the plate, in a line with the cuspid and space. The plate being in position, a rubber band was passed over the spur and drawn tight to the ring on the tooth by means of a ligature, the band in its course resting in the notch of the elevation on the plate. By this arrangement no pain was inflicted except that incident to forcing the ligature into position under the gum, while power was exerted in a nearly direct line with the long axis of the tooth, and in a gentle, continuous manner.

Another and most excellent plan of securing attachment to a partially erupted cuspid, is that recommended by Prof.

J. F. Flagg. It consists in screwing a gold ring-bolt or screw-eye into the point of the cusp. The screw-eye can be made by soldering a small gold ring to the end of a section of threaded gold wire. After the correction is accomplished, the screw is removed and the hole filled with gold.

A very simple and effective appliance, securely attached to the teeth, for drawing a cuspid down into position has been devised by Dr. Angle, and is shown in Fig. 67.

As will be seen, the bicuspid is fitted with a metal band to which is soldered a short piece of tubing. A wire of suitable length is flattened at one end and bent into a hook to engage with the cutting edge of the lateral, while the other end is bent at a right angle to fit into the tube on the bicuspid band. Midway of the length of this wire is soldered a small button. The unerupted cuspid has a headed pin cemented into its labial surface or point of cusp, and over this pin and the button on the wire is stretched a section of rubber tubing to produce the desired tension.

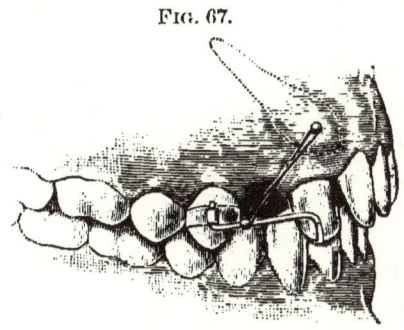

FIG. 67.
Drawing down Cuspid. (Angle.)

If mal-position of an erupting cuspid should be complicated with more or less torsion, the correction of the latter will be best accomplished after the tooth is nearly or quite in position.

CHAPTER III.

INCISOR TEETH SITUATED OUTSIDE OR INSIDE OF THE ARCH AFTER DENTITION IS COMPLETE.

Irregularities of this character will require much the same treatment as similar cases occurring during dentition, but the attendant difficulties will be greater, owing to the increased density of the alveolar structure and the presence of all the teeth, making the obtaining of space more difficult. In the lower jaw, the irregularity in most cases is confined to one or two teeth, standing either anteriorly or posteriorly to the line of the arch. If they are located posteriorly, and the extraction of one of them be not indicated, room should be made (if it does not exist) by pressing apart the neighboring teeth. After this is done, they may be conveniently forced into place by means of a Coffin plate, constructed as shown in Fig. 68.

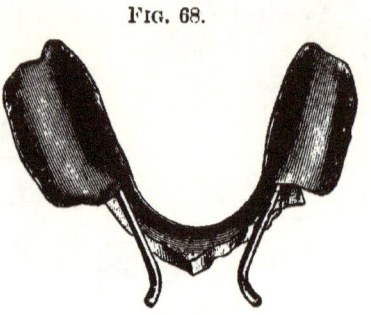

FIG. 68.

Coffin Plate for Lower Incisors.

When a single lower incisor is locked inside of the arch by the over-lapping of its neighbors, it is often so firmly held in its mal-position that all ordinary means will fail to move it unless space is first provided for it by lateral pressure. This being sometimes difficult of accomplishment, the direct power of the jack-screw may be taken advantage of in such cases to overcome the difficulty, as shown in Fig. 69.

A platinum band was constructed to fit the lateral, and on its lingual surface was soldered a tongue of heavy platinum, so formed that it would lie in contact with the tooth when the band was in position. Into this tongue, near its free end, was drilled a counter-sunk hole nearly deep enough to pass through the metal. On the opposite side of the mouth the second bicuspid was similarly fitted with a band, to which was soldered a strip of platinized gold long enough to cover the lingual surface of the adjoining molar. By this means the molar was made to assist in resisting the force to be applied to the lateral. The bicuspid band was also re-enforced by an additional piece of heavy platinum soldered to it at a point diagonally opposite to the lateral. Into this latter piece a horizontal slot was drilled with an engine-bur, sufficiently deep and long to receive the fish-tail end of an ordinary nickeled-steel jack-screw. After both bands were cemented in place the jack-screw was placed between them, with the flat end in the bicuspid band and the point resting in the counter-sunk hole of the lateral band. The patient increased the tension of the screw from day to day by turning, and in two weeks' time the tooth was in line. It was held there until it became firm by means of platinum binding wire woven about it and its neighbors.

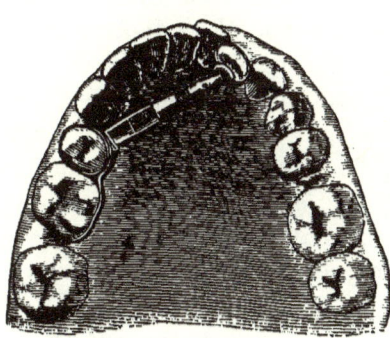

Fig. 69.

Jack-Screw Forcing Out Inferior Lateral.

In cases where it is not deemed advisable to pursue the plan just mentioned, an excellent way of creating space and at the same time moving an incisor outward into line is by the employment of a double-acting device, composed of a thin metallic ribbon and spring, or bolt and nut.

The first recorded suggestion of an appliance of this character appears in one of Dr. Farrar's articles, published in 1884.*

---

* *Dental Cosmos*, Vol. XXVI., p. 672.

Prof. Angle employs a modified and simplified device, as is shown in the accompanying illustration. It is constructed as follows: The ribbon being of sufficient length to pass back of the inlocked tooth and rest slightly upon the labial surfaces of the adjoining teeth, two short tubes are soldered to it, one at each end. One of these tubes is set vertically and the other horizontally. A piece of steel wire, bent at a right angle at one end and thread-cut and provided with a nut at the other, is made to engage with the tubes, the bent end slipping into the vertical tube and the other passing into the horizontal one, with the nut resting against its inner end. By unscrewing the nut, the ends of the ribbon are forced apart and the desired movements accomplished. Fig. 70 represents the appliance in position, and Fig. 71 the separate parts of which it is constructed. In this device the direct power of the screw is used to furnish the necessary pressure.

FIG. 70.

FIG. 71.
(Angle.) 

Instead of the nut and bolt, Prof. Matteson prefers a coiled-wire spring to operate upon the ends of the ribbon, as shown in Fig. 72.

The spring is made from piano-wire, No. 14 or 16, and when in place the ends rest in two short tubes soldered horizontally to the ribbon near its extremities. The tubes have slots cut into their upper surfaces to prevent the spring from pressing upon the gum.

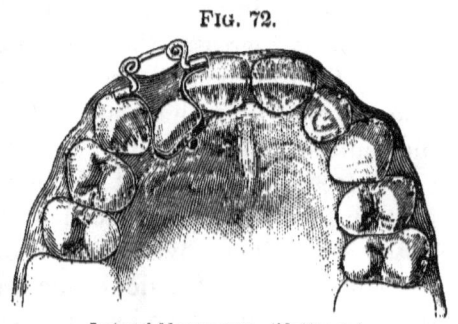

FIG. 72.

Lateral Movement. (Matteson.)

Should the tension of the spring not be sufficient to move the tooth entirely into place, a longer one may be substituted for it.

As will be noticed, the ribbon has a short-headed pin or post soldered to its exposed surface opposite the centre of the inlocked tooth to furnish a ready means of ligating the band to the tooth should it be necessary to prevent its slipping out of position.

When an incisor tooth in the lower jaw stands outside of the arch, the malposition is usually due either to its having been forced out of place by a superior one occluding back of it, or to unusual crowding on the part of its neighbors. In the first instance, the correction of the occlusion of the superior tooth will usually press the lower one into its proper place, while in the second instance, it will be necessary to consider the advisability of extracting one of the crowded teeth to afford room. If such extraction be deemed best the case will be greatly simplified and the malposed tooth can be brought into line by some one of the means described for drawing inward the superior incisors.

If it be deemed inexpedient to extract one of the crowded teeth, room will have to be provided either by expanding the arch or by extracting a tooth or teeth back of the cuspids.

In considering the matter of expansion of the arch it should be borne in mind that the enlargement of one arch may also necessitate the expansion of the other in order to preserve the normal occlusion. If both jaws will admit of it to advantage, it may be the best plan to pursue, although it will necessarily increase the labor and difficulty of the operation. Generally, if the occlusion and facial expression be satisfactory, it will be far better not to disturb the general relation of the teeth, but rather to extract one or more of the bicuspids or molars. After any of the posterior teeth have been extracted, the anterior ones can be moved apart or backward and the irregular tooth brought into place.

The inferior incisors, after being brought into line, will usually be retained in place by the occlusion of the superior teeth, but where this is not the case, they may be retained by means of platinum binding wire woven about all of the incisors at or near their necks, or they may be securely held by means of a ribbon of thin gold fitting the lingual surfaces of the incisors, to which is soldered a platinum band to encircle each tooth that has been corrected. The piece is set with phosphate of zinc as a lining to the bands.

For drawing or forcing into line any of the superior incisors standing outside of the arch, a variety of methods is at our disposal. In the upper jaw the extraction of one or more incisors to provide room for other outstanding ones is, except in rare cases, not to be thought of, although, as just stated, in the lower jaw extraction may often be advantageously resorted to. The greater conspicuousness of the superior incisors, and the difference in size between the centrals and laterals would cause the absence of any one of them to be most noticeable. Rare cases occur, however, in which such extraction is justifiable, as already described, but a wise discrimination must be exercised in regard to the matter, as otherwise a greater deformity is likely to be created than the one already existing. Where space is needed in the arch for the outstanding tooth or teeth and expansion of the arch is not indicated, we may obtain it by extraction back of the cuspids, or where the lack of space is slight in amount it may be secured by simply exerting pressure upon the adjoining anterior teeth. A simple way of producing this pressure is by the use of compressed wood, as described on page 78.

Another plan is by means of a vulcanite plate to which are attached gold or steel wires so arranged that their free ends when drawn together and inserted in the space intended to be widened, will press the adjoining teeth farther apart.

Still another, without the use of a plate, which the author has found very effective, consisting of platinum bands attached to the teeth to be moved, with a piano-wire spring acting between them, is described and illustrated in part III, chapter VII.

Dr. Farrar recommends for the same purpose a delicate jack-screw with crutch ends to fit the teeth to be separated.

Prof. Goddard employs for the same purpose an appliance, as shown in Fig. 73, and constructed as follows: —

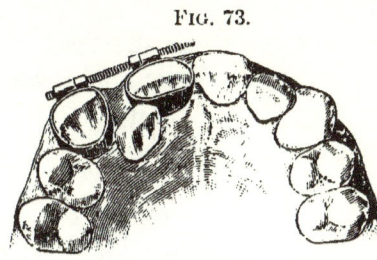

Fig. 73.

Appliance for Separation. (Goddard.)

The two teeth bordering the space are encircled by bands having short open tubes soldered to their labial surfaces in a horizontal position. Through these tubes is passed a threaded wire having two nuts upon it. One of these is designed to simply offer resistance, while the other, by being turned, will gradually force the teeth apart.

After the desired space has been obtained, the inlocked tooth may be brought into place by some one of the methods about to be described.

In devising appliances for moving the superior incisor teeth either inward or outward into line, due consideration must be given to the occlusion. To avoid conspicuousness, it is desirable to have the operating appliances placed within the arch, but very frequently the occlusion of the lower teeth will interfere with such arrangement. In the latter case they may be so constructed as to operate from the outside.

One of the simplest methods for moving one or more incisors outward into line is by the employment of the Coffin solid plate, as shown in Fig. 48. The only difficulty met with by the author in the use of this form of plate has been where the teeth to be moved, although inside of the arch, stand perpendicularly or incline slightly forward. In these cases the free ends of the wires, after being pressed up into position on the teeth, are frequently thrown down toward the cutting edge by the force of the spring operating upon an inclined surface. When great inconvenience arises from this cause, it may be remedied by cementing a narrow platinum band about midway of the crown of the tooth to be moved, and placing the end of the wire spring above it.

Another plan for moving outward any or all of the superior incisors, is by means of a plate constructed as shown in Fig. 74.

A thin vulcanite plate is made to cover the roof of the mouth and cap the bicuspids and molars; opposite the tooth or teeth to be moved, the plate is allowed to come nearly down to their cutting edges. Directly opposite to the center of each of these teeth, a hole is drilled entirely through the rubber to receive a piece of screw wire long enough to pass through and project a little beyond the plate. In springing the plate into position the slightly projecting ends of the screws will press against the teeth and they will be moved forward. A half turn of the screws every day will soon force the teeth into position.

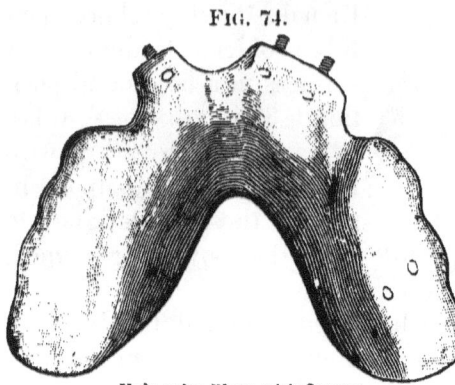

Fig. 74.
Vulcanite Plate with Screws.

Dr. Dodge * suggests the employment of a hollow metal screw tipped with gutta-percha at its exposed end used in connection with a vulcanite plate, as just described,

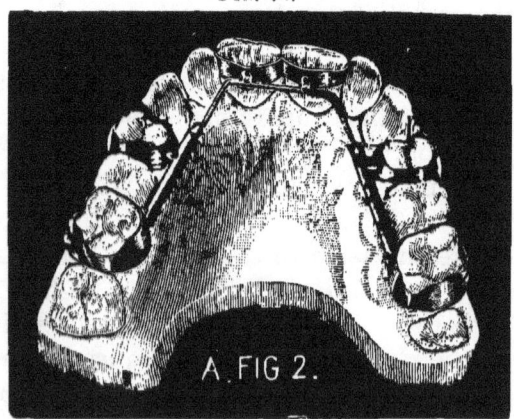

Fig. 75.
Moving Centrals Outward. (Matteson.)

---
* *Dental Cosmos*, Vol. XXXI., p. 772.

claiming for it greater friction in contact with the tooth and non-liability to injury of tooth substance.

Less cumbersome than rubber plates and more positive and satisfactory in the majority of cases are appliances constructed entirely of metal.

Fig. 75 shows one of this character designed by Prof. Matteson,* somewhat on the Angle plan, for the purpose of moving forward two inlocked superior central incisors. As will be seen, it is firmly attached to the anchor teeth by cemented bands and is operated by turning the nuts which rest against tubes soldered to the anchor bands.

Should the occlusion not permit the use of the appliance just described, the same end may be attained by employing a fixture devised by Dr. Kirk † and illustrated in Fig. 76.

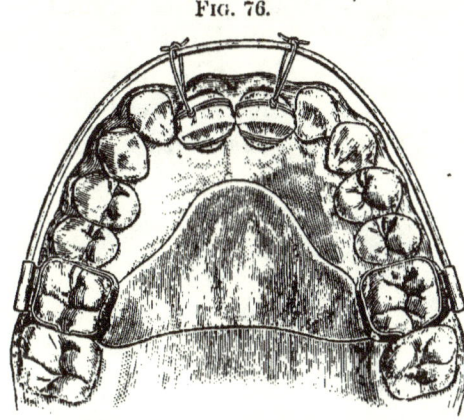

Fig. 76.
Plate, Band and Bar Appliance. (Kirk.)

It consists of a narrow silver plate swaged to fit and partly cover the roof of the mouth, to which are attached two broad clasps of platinized gold fitting the first molars. To the buccal surfaces of these clasps are soldered tubes closed at their distal ends to receive a gilded piano-wire, bent to conform to the outline of the arch, but slightly longer.

When in position, the inlocked centrals are tightly ligated to the wire immediately in front of them, which by its elasticity draws them forward.

---

* *Dental Review*, July, '92, p. 564.
† *Dental Cosmos*, Vol. XXXIII., p. 908.

Another appliance, simple in construction and not interfering with occlusion, designed to draw one central outward into line and at the same time press the adjoining prominent one back into place, is shown in Fig. 77.

It was devised by Dr. Jackson and is constructed after his method. The first molar is fitted with a crib to which the extending spring wire is attached. This engages at its free end with a tubed band cemented to the inlying central, and in its course rests upon and presses against the prominent central. A double movement is thus produced and the power of the spring may be increased as desired by straightening its curve.

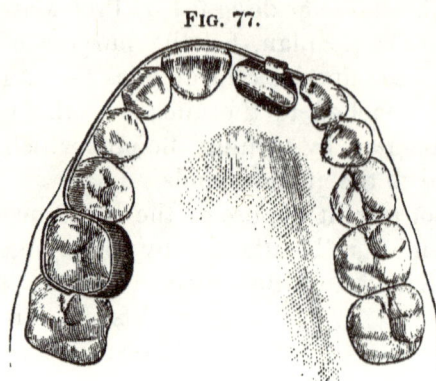

Crib, Band and Spring Devise. (Jackson.)

Still another device, even more simple than the preceding one, for moving forward an inlocked incisor, is illustrated in Fig. 78.

It also is one of Dr. Jackson's, and consists simply of a tubed-band attached to the malposed tooth and an ingeniously arranged wire spring to furnish the motive power.

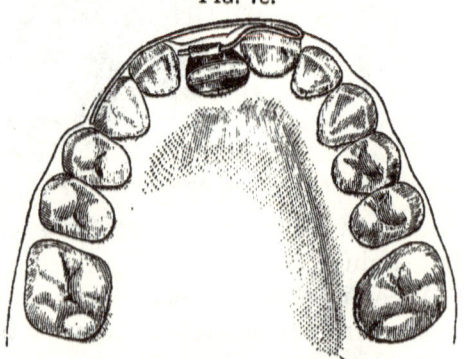

Tubed-Band and Spring. (Jackson.)

The spring is formed by bending a piece of piano-wire

into the form of a loop with one end much longer than the other, and both of them suitably curved to follow the outline of the arch. The longer arm of the spring should at least be long enough to cover the surfaces of three teeth to furnish proper support.

In adjusting the spring, the longer arm should be next to the gum while the shorter one is being inserted into the tube, then by turning it downward the whole appliance comes into proper position with the shorter arm acting as a spring to draw the incisor outward.

When the tooth is in place it may be retained by inserting a short wire into the tube and allowing it to rest upon the labial surfaces of the two adjacent teeth.

When sufficient time has been allowed for the tooth to become firm, (never less than six months) the retainer should be carefully removed. For a few months afterward the patient should be seen once a week, in order to ascertain whether the tooth is remaining in its new position. Should it manifest a tendency to recede, the retainer must again be placed in position and kept there for a further period of three months or more.

By thus carefully watching a case after its supposed completion, we may often avoid the loss of some of the ground we have gained.

CHAPTER IV.

CUSPID TEETH SITUATED OUTSIDE OR INSIDE OF THE ARCH.

Of the various forms of irregularity that present for treatment, none perhaps is more common than that in which the cuspid teeth are located outside of the arch. The cause most frequently responsible for this condition is the premature extraction of the temporary cuspids, although it is often caused by delayed eruption of the permanent ones, and by the lack of accommodation a small arch sometimes affords for the full complement of teeth. The cuspids (superior) being among the later teeth to appear, often find their territory pre-occupied by the earlier arrivals. Frequently, though not always, the mal-position of the cuspids is associated with like mal-position of certain neighbors, usually the central and lateral incisors. The irregularity of these adjoining teeth is, in most cases, brought about by the pressure of the cuspids in their attempt to occupy their places; for, previous to their appearance there is no inducement, if the occlusion be normal, for the incisors to vary much from their true positions. The fact should not be overlooked that all teeth in erupting are impelled by a strong hidden force to seek their proper positions in the line of the arch, and in no teeth is this persistence more plainly or powerfully exhibited than in the cuspids.

The conditions being favorable each tooth will naturally assume its place in line, and should obstructions interfere it will strive to overcome them; but the cuspid teeth will, if necessary, exert a power far exceeding that of any of the other teeth in their efforts to gain their proper positions in the arch. To this end incisors are often disarranged, and bicuspids forced inward or outward. This wonderful force exerted by the cuspids, may well be illustrated by a case which occurred in the practice of the author many years ago:

The patient was a young lady about fifteen years of age, in whose upper jaw a cuspid had erupted outside of the arch, causing projection of the lip. All of the other teeth were regular, but the bicuspids and molars on the affected side were somewhat in advance of their true positions, and there was consequently very little space in the arch for the accommodation of this cuspid. The first molar on the same side was badly decayed, so it was decided to extract it preliminary to making room for the cuspid. An appliance was then attached to the second molar and second bicuspid, intended to draw the latter tooth backward. The patient left with this fixture in position and did not return until eighteen months later, when it was noticed that both bicuspids had moved backward and the cuspid occupied its normal position in the arch. It transpired that the appliance, having caused some pain, was removed by the patient two days after it had been placed in position. The correction of the irregularity had been entirely accomplished by the cuspid forcing its way into place and crowding the bicuspids backward in the effort.

To obtain space for the accommodation of the cuspid when it is situated outside of the arch, we usually have to decide between the enlargement of the arch and the extraction of a tooth anterior or posterior to it. If the upper arch is contracted and will admit of expansion to advantage, it may be done by one of the methods described in Chapter VII. of this part; but if this be not indicated, we will have to decide upon the extraction of a bicuspid or lateral in order to obtain space.

A careful consideration of the rules governing extraction, Part I., Chapter VII., will greatly assist the operator in deciding which tooth to extract.

It very frequently happens that the space in the arch intended to accommodate the cuspid is nearly, but not quite, sufficient. In such cases, slight additional space may generally be gained by pressing apart the adjoining teeth with the fixture shown in Fig. 123.

Room having been provided, the cuspid tooth may be brought into place by one of several methods that are equally effective in the upper and lower jaws. Outstanding cuspids are usually situated a little in advance of their normal positions, so that in bringing them to place we must exert force in a backward as well as inward direction.

Where from the appearance of the teeth and the surrounding parts it seems probable that great force will not be required, a cuspid may frequently be drawn inward by so simple a means as that shown in Fig. 79.

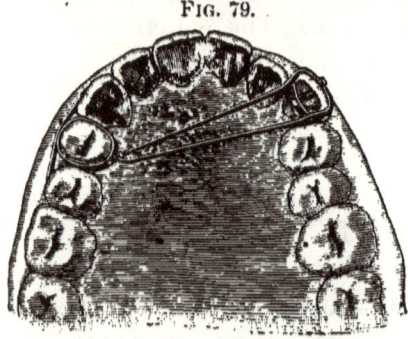

Fig. 79.
Metal Bands and Rubber Ring.

In this case a platinum band, with a pin on its labial face, was cemented to the outstanding cuspid. To the first bicuspid on the opposite side was fitted a similar band with a small gold hook on the palatine surface and a bar of platinized gold on the buccal surface long enough to extend to and rest upon the adjoining cuspid and second bicuspid. This provided the resistance of three teeth, whilst attachment was made to but one. A light vulcanite plate was made to cover the arch, so as to protect it from the irritation of the rubber ring, which was stretched from band to band. The operation of bringing the tooth into line was somewhat slow, occupying some four or five weeks' time, but the object was satisfactorily accomplished.

Where the movement to be effected is more backward than inward, it may often be very satisfactorily and easily accomplished by the simple appliance shown in Fig. 80.

A platinum band, with short gold wires soldered to the buccal and lingual surfaces, is cemented to the tooth to be moved, while a similar one is attached to a molar or other anchor tooth. The wires on the anterior band are bent forward, and those on the posterior one are curved backward. Two rubber rings, caught over the gold hooks, connect the two bands and yield the tractile power required. These rubber rings can be removed and replaced for cleansing the teeth, or can be renewed at will by the patient. Two rings can be attached to each pair of hooks, if greater power be required, or the same object can be attained by cutting wider rings from thicker tubing.

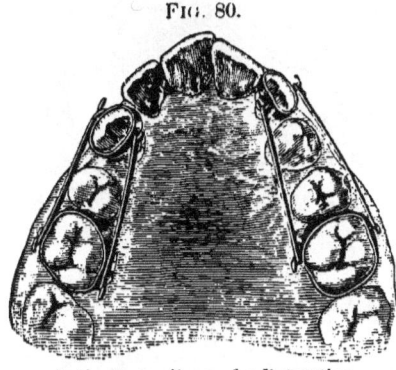
Fig. 80.
Author's Appliance for Retraction.

Another simple way of bringing about the same movement, is by means of the Coffin plate with the wire or wires attached to the buccal portion and extending forward until their free ends rest upon the teeth to be moved. Ordinarily, the pressure to be exerted by them would be inward only; but by bending their ends into the form of partial hooks, so as to engage with the mesial surfaces of the teeth, an additional backward pressure is obtained.

In most cases, however, greater force than that exerted by a rubber band or spring wire will be necessary to draw a cuspid into place, especially if it be large and firmly implanted. In such event the power exerted by a screw in some form will probably yield the best results.

One of the simplest and best appliances for drawing a cuspid backward and inward into line is that devised by Prof. Angle and shown in Fig. 81.

The first molar is encircled by a metal band, to which

on its palatal surface is soldered a long piece of tubing to accommodate the traction screw.

The cuspid is also encircled by a band with a short tube soldered horizontally to it on its distal surface with which the bent end of the traction screw engages.

The nut operating against the distal end of the long tube will rapidly move the cuspid into position.

FIG. 81.

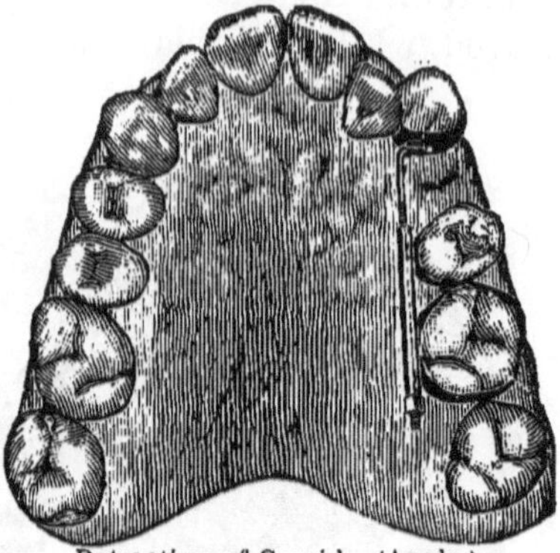

Retraction of Cuspid. (Angle.)

Fig. 82 illustrates another appliance of Prof. Angle's, very similar in character, but with the tube and screw located upon the outside of the arch.

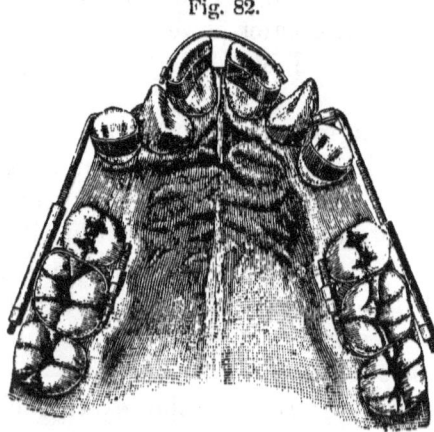

Fig. 82.

Backward Movement and Rotation. (Angle.)

It will be noticed that on the left side the short pipe or tube is attached to the cuspid band at the mesio-buccal angle of the tooth in order to rotate it as well as draw it backward, while upon the right side the screw is hooked over a spur upon the cuspid band to accomplish the same purpose more conveniently.

Dr. Farrar's device for effecting the same movement is shown in Fig. 83.

It consists of a narrow ribbon of gold, long enough to enclose the cuspid tooth and some tooth back of the space it is to occupy. The ends of this ribbon nearly meet on the buccal side of the teeth, and after being reënforced with studs of heavy gold, the anterior one being simply drilled and the posterior one drilled and threaded, they are connected by means of a gold screw. The turning of the screw brings the ribbon ends nearer together, and causes corresponding traction on the misplaced tooth. The ribbon, at suitable places, has ears or tips attached to it, intended to rest upon the masticating or inclined surfaces of the enclosed teeth and prevent the band from slipping up and irritating the gum.

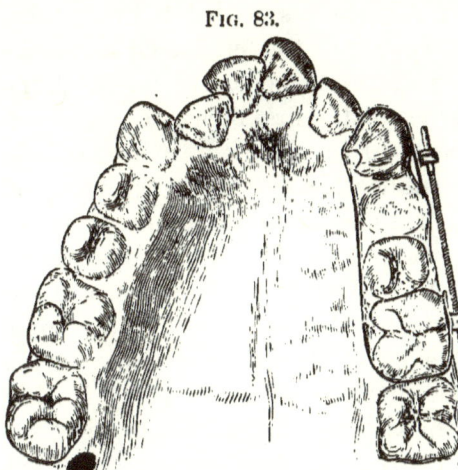

Fig. 83.

Farrar's Traction Apparatus.

Prof. E. T. Darby's plan for producing the same movement is by the use of a rubber plate, a gold encasement for the cuspid and a gold screw for connecting the two and producing the required tension. Fig. 84 is drawn from one of his models, and represents the fixture in position. The case was that of a young lady, fourteen years of age, who applied for the correction of irregularity of the anterior teeth. As will be

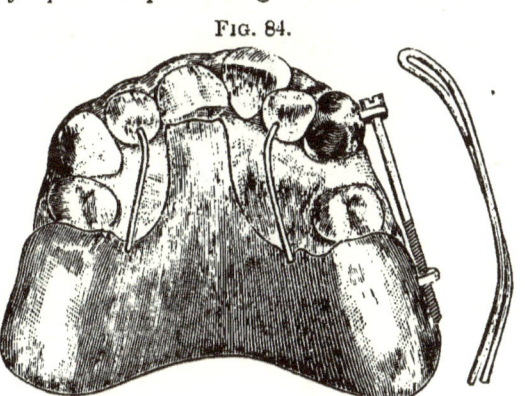

Fig. 84.

Darby's Appliance for Retraction.

noticed in the illustration, both laterals and the right central were inside the proper line of the arch, while the left central was outside of it. Space was needed to bring these teeth into position, and to obtain it the left cuspid had to be moved backward in the arch. Opportunity for so doing was afforded by the absence of the first bicuspid.

To move the cuspid backward, and to assist in accomplishing other movements, a rubber plate covering the arch and capping the molar teeth was constructed, and into it on the buccal surface was inserted a gold stud or ear, drilled and tapped. A gold helmet to cover the entire crown of the cuspid was then constructed, with a projection on the labial surface drilled for the passage of the traction screw. After this helmet was cemented in place with phosphate of zinc, and the plate inserted, the two were connected by means of a long gold screw. Twice each day this screw was turned, until the cuspid was brought almost into contact with the second bicuspid.

While this movement was progressing, other objects were being accomplished. The rubber plate when first inserted had a piano-wire spring attached to its palatine surface to force forward the right central. This accomplished, the spring was removed and rubber added to the plate, to keep this tooth in its new position. Two new piano-wire springs were next inserted to spread apart and press forward the laterals, as shown in cut. They were brought into position by the time the cuspid had been drawn sufficiently backward.

The helmet and screw were now removed and a piece of piano-wire, doubled and bent to proper shape, was inserted in the hole of the gold stud in the rubber plate, in such a way that the end would rest upon the outstanding central and force it into line.

The case as corrected is shown in Fig. 85. The entire work of correction, with its varied movements, occupied but five months' time, and was accomplished by the use of a single plate with its different attachments. To retain the teeth in position a rubber plate was worn covering the arch and having a gold T inserted to pass between the centrals.

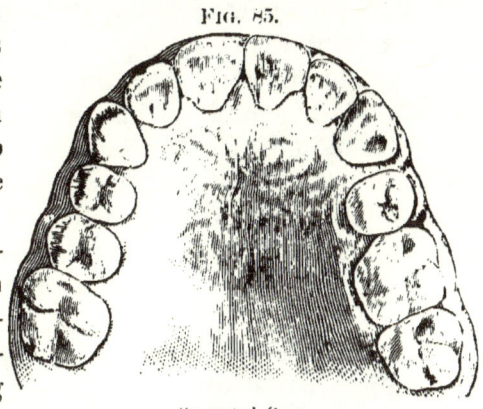

Fig. 85.

Corrected Case.

Where the occlusion of the teeth would not contraindicate its employment, an outstanding cuspid may be drawn inward by means of a screw operating between the tooth to be moved and those used as anchorages.

Fig. 86 represents a case of this character, where, in addition to the firmness of the tooth, the patient resided at such a distance from the dentist that a visit to him could be made only at intervals of two or three weeks. It was therefore necessary to devise an appliance of such character that it could not be removed or misplaced, and with a sufficiency of power that might be regulated by the patient herself. The appliance shown in cut, consists of two platinum bands made to fit the misplaced cuspid and opposite molar respectively, and cemented to these teeth. To the palatine surface of each of these bands was soldered a gold ring, which served as point of attachment for the gold box and screw, which operated between them.

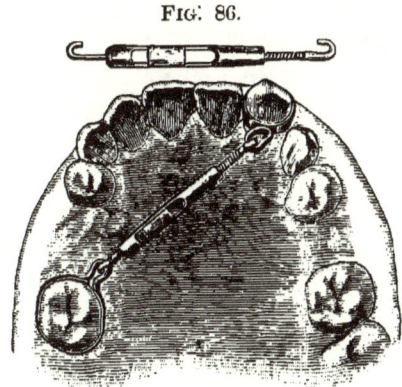

Fig. 86.

Gold Box and Screw Drawing in Cuspid.

One end of the gold box was bushed and thread-cut to receive the gold screw, which at the opposite end was bent into the form of a hook to engage with the ring on the cuspid band. The other end of the box was fitted with a smooth gold wire, with a head on one end to serve as a swivel, and a hook on the other to attach to the ring on the molar band. Turning the box with a wrench drew the screw inward, and with it the cuspid tooth. Using a single molar for anchorage in the movement of a cuspid was scarcely in accord with correct practice, but in this case there was no alternative. In drawing the cuspid to place the molar was also moved somewhat inward and forward, but it soon resumed its former position after being relieved from duty. The corrected tooth was retained in place by having cemented to it the small band and bar appliance shown in position and separately in Fig. 87.

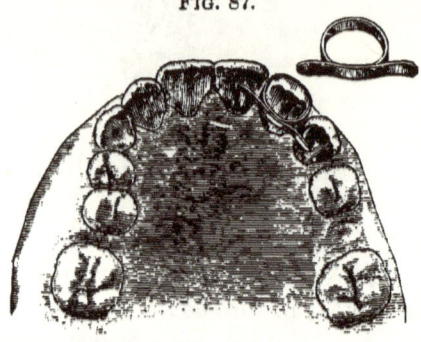

Fig. 87. Completed Case with Retaining Appliance.

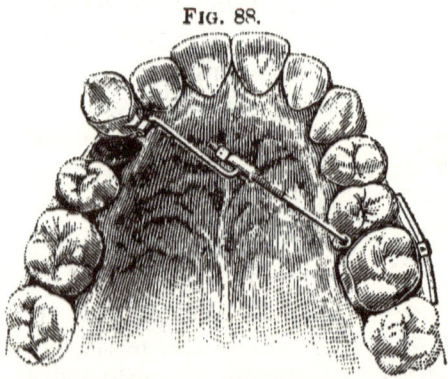

Fig. 88. Combination Appliance for Drawing in Cuspid. (Angle.)

Another appliance of the same general character, but different in construction, and designed by Prof. Angle, is illustrated in Fig. 88. It is described as follows:—

"The cuspid tooth is banded and a piece of gold wire, bent sharply at right angles, is hooked into a pipe soldered to the surface. The other end of the wire is soldered to a pipe through which the small traction screw is slipped, and against which the nut works.

"The other end of the traction screw is hooked into a pipe, soldered to a band, encircling the first molar. The anchorage of this tooth is further reinforced by a piece of the gold wire, which is slipped through a tube soldered to the buccal surface of this band, the end of the wire resting against the adjoining teeth.

"If the tongue becomes abraded by the end of the screw as it emerges from the nut, a very nice way of protecting it, as in all similar cases, is for the patient to lay over the end of the screw a very small piece of the very common article known as chewing gum."

Occasionally it is possible to move a cuspid inward and at the same time provide room for its accommodation by pressing the adjoining teeth apart.

Fig. 89 shows an appliance for this purpose devised and used successfully by Prof. Goddard.

It was designed to operate upon the principle of a double wedge, the cuspid serving as one wedge and the V-shaped strip of metal as the other. A nut and bolt operating between the two, as shown, furnished the motive power. The strip was altered in form as the work progressed, always, however, retaining its wedge shape.

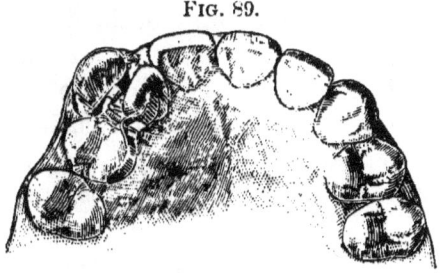

FIG. 89.

Appliance for Drawing in Cuspid. (Goddard.)

Prof. Matteson * illustrates and describes the use of a novel, but simple, fixture for producing the double movement previously described. It is shown in position in Figs. 90 and 91. It consists of two flexible metal strips attached to and held apart at their inner ends by a suitably-shaped

---

* *Dental Cosmos*, Vol. XXXIV., p. 247.

wire made long enough to rest upon the palatal surfaces of the two teeth bordering the space to be occupied by the cuspid. At their outer extremities these metal bands were arranged to engage with a curved bolt and nut overlying

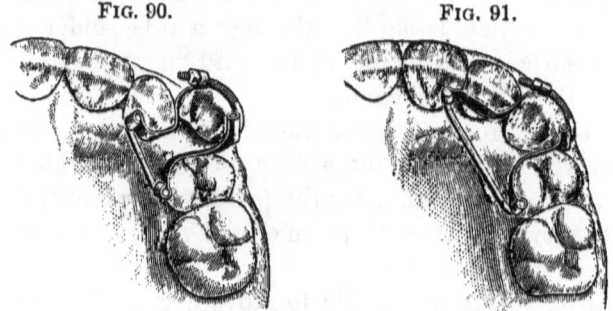

FIG. 90.   FIG. 91.

Creating Space and Moving in Cuspid. (Matteson.)

the outstanding cuspid in such a way that when in position, as shown, the turning of the nut would draw the ends of the strips towards one another and thus force the cuspid inward at the same time that the adjoining teeth were forced apart to provide accommodation for it.

Fig. 90 shows the appliance as first used and Fig. 91 the same with a longer wire substituted for the shorter one after the latter was rendered unserviceable by the moving of the tooth.

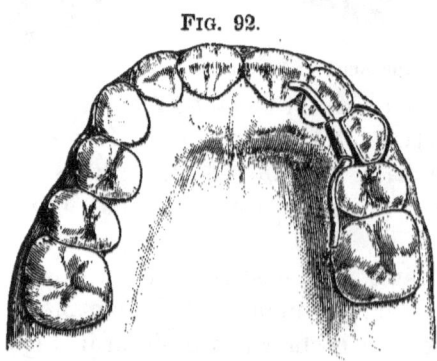

FIG. 92.

Retainer. (Matteson.)

In the case in hand, after the preliminary wedging, only two weeks' time was consumed in bringing the cuspid into place with the appliance, as described, although the patient was a well-developed man, twenty-two years of age.

The tooth was retained in place by means of a tubed band cemented to it with wire inserted to rest against adjoining teeth, as shown in Fig. 92.

When a superior cuspid erupts inside of the arch the difficulties attending its being brought into position are far greater than when it erupts externally. This is partly due to the fact that the space between it and the opposite side of the arch is too limited to admit of the use of some of our best power-yielding appliances and partly, also, to the thickness of the alveolar process in which it is embedded and that will have to be resorbed before the tooth can assume its proper position.

The power to be applied to an in-lying cuspid must necessarily be very great to carry with it any prospect of success. A solid Coffin plate, with a very stiff piano-wire embedded in it, will yield the greatest amount of spring power, but where this is insufficient we must needs resort to the jack-screw in some of its forms.

An appliance of Dr. Angle's, as shown in Fig. 93, for forcing outward a cuspid is neat, simple and effective.

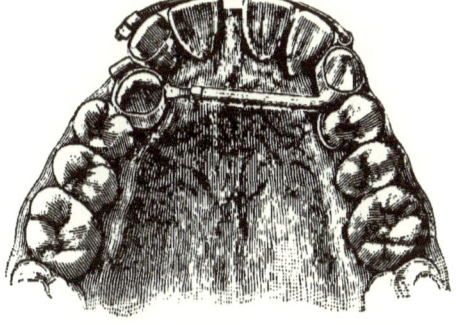

Fig. 93.

Jack-Screw Moving Out Cuspid. (Angle.)

"The base of the jack-screw is soldered to a band encircling the opposite cuspid and reinforced by a spur resting against the first bicuspid, and also by the large traction screw which is hooked into a pipe soldered to the labial surface of the band and passing in front of the incisors through a tube soldered to a band on the labial surface of the lateral incisor, against which the nut works.

"In this case, the left central and lateral were moved forward in the line of the arch, thereby closing the space between the centrals, and, at the same time, providing space for the out-moving cuspid. The large screw was beaten flat and polished before insertion."

CHAPTER V.

MISPLACED BICUSPIDS.

The bicuspid teeth, both superior and inferior, are often found located outside or inside of the normal arch line, but their mal-position is not of as frequent occurrence as that of the anterior teeth.

Their position out of line, as in the case of most forms of individual irregularity, is due to lack of space or the crowding of other teeth. Sometimes, through tardy eruption, their space in the arch has been encroached upon by the pressure of the erupting cuspids in front, as well as the forward-moving tendency of the molars. In such cases one or both of the bicuspids are compelled to assume a position outside or inside of the arch, the latter being the one they most commonly take.

Again, their predecessors, the deciduous molars, frequently have their crowns destroyed by caries long before the time for their natural removal, while their roots remain. Inducement is thus offered for the adjoining teeth to occupy part of the space, and the bicuspids are forced to erupt in an abnormal position.

In other cases, they may have taken their places in line, or nearly so, and are subsequently forced out of place by the effort of the cuspids to occupy their places in the arch. The ease with which they may be forced out of position is readily understood when we consider that their roots are conical and rather short, and that they are placed between teeth that are firmly set and have either a single long root firmly implanted, like the cuspids, or several roots, like the molars. Their distinctly convex approximal surfaces also greatly favor their displacement.

The second bicuspid is more frequently found out of line than the first, probably because of its later eruption.

The lack of alignment of one or both bicuspids is sometimes associated with a greater or less degree of torsion; but this is not of common occurrence, and when met with is either corrected in the act of bringing the tooth into line or will have to be remedied by a separate operation afterward.

The greater or less difficulty of bringing into line one or more bicuspids situated inside of the arch, will usually be entirely dependent upon the amount of space existing for their accommodation. If much of their space in the arch has been pre-occupied by adjacent teeth, these will first have to be pressed apart to afford accommodation. Should full or nearly full space exist for them in the arch, they may usually be forced into line by the elasticity of a vulcanite plate, or of metal in some form of spring. Where it is designed that the moving tooth shall make room for itself as it advances, the greater power of the jack-screw will be required.

A simple method of moving a bicuspid, either upper or lower, outward into line is to obtain a plaster model of the jaw. The plaster tooth representing the one to be moved should then be cut away on its palatine or lingual surface until this portion of it is in line with the same surfaces of the adjoining teeth. A vulcanite plate made upon this model with a piece of piano-wire embedded in its central portion, if it be for the lower jaw, will, by its elasticity, soon bring the tooth into position. Or, we may make the plate upon the unaltered model and then insert a wooden peg in a hole drilled in the plate opposite the tooth to be operated upon. Or, instead of the wooden peg, a metal screw may be inserted so as to act upon the tooth. By setting the screw well into the rubber plate, it may be elongated by unscrewing from time to time until the object is attained.

Dr. Talbot has devised an excellent method of forcing one or more bicuspids into line by means of a coiled spring of piano-wire, in connection with a rubber plate to hold it in position and properly direct its action. Fig. 94 represents the appliance in position. Dr. Talbot says: * "A thin, narrow, close-fitting, vulcanite plate was made, and a hole drilled through the middle of it opposite the centre of the tooth to be moved. In the other side, another hole was drilled, but not quite through the plate. A suitable spring, Fig. 95, was then made of piano-wire, having a single coil A, and the ends of its arms bent at about a right angle. One of these ends, C, was cut short to enter the corresponding hole in the plate, and the other end, B, left long enough to go through the plate and impinge on the lingual surface of the bicuspid, leaving a full eighth of an inch between that arm of the spring and the plate, as is clearly shown by Fig. 94, where the spring is in position to act upon the tooth to be moved. Both the spring and the plate may be removed instantly, either for cleasing purposes or to increase the power of the spring by spreading its arms, or to open the coil so that the tooth may be held steady at the point to which it has been moved. Fig. 96 shows a spring having two

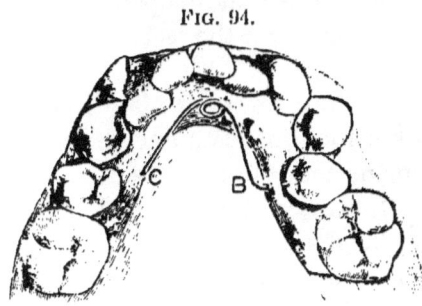

Fig. 94.

Talbot's Vulcanite Plate and Coiled Spring.

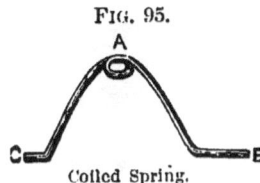

Fig. 95.

Coiled Spring.

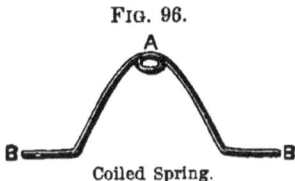

Fig. 96.

Coiled Spring.

---

* *Dental Cosmos*, Vol. XXVIII., pp. 286-7.

long ends, B B, which is designed for a case in which two such teeth are to be moved in opposite directions."

The advantage of this appliance is that it operates without occupying any of the space between the teeth, which in most cases is important.

Where there is no great crowding, however, Magill bands may be attached to the anchor tooth or teeth and the one to be moved, and the Talbot spring made to rest in suitable depressions formed in them. In this way the objection to a removable rubber plate may be done away with.

Where the superior power of the jack-screw is to be taken advantage of, Dr. Kingsley's method of using it in combination with a slotted vulcanite plate is one of the best.

The accompanying illustrations, Figs. 97 and 98, copied

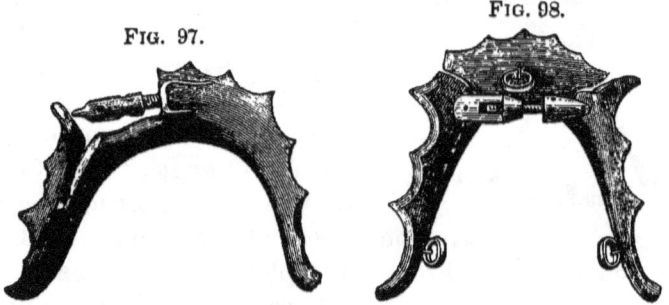

Kingsley's Slotted Vulcanite Plates with Jack-Screw.

from Dr. Kingsley's work,* represent some of the ways in which he accomplishes movements, slightly varying in character. Fig. 97 operated to move outward both bicuspids of the left side inferior, the first more than the second; while Fig. 98 moved all four of the inferior bicuspids.

Where it is desired to avoid the use of a plate, Magill bands, re-enforced, drilled and counter-sunk, may be cemented to the teeth to be moved and the jack-screw inserted between them. Prof. Angle's device for expanding the arch, as shown and described on page 106, may also be

* Loc. cit.

advantageously used for moving outward one or more of the bicuspids. It will be noticed that in the operation of this appliance any instanding teeth are moved outward into line before real expansion of the arch begins; if, therefore, the moving of individual teeth is alone desired, operations can be suspended as soon as that object has been accomplished.

The small size of the jack-screw in the Angle device is also an element in its favor, since it will interfere less with the movements of the tongue than the larger ones commonly used.

In addition to the power of the jack-screw, it has the further advantage of rapidity of action; so that, if its position in the mouth should somewhat inconvenience the patient, it would do so only for a very short time.

Dr. Jackson has very ingeniously adapted his spring and crib method to the moving of bicuspids, as shown in the following illustrations:—

Fig. 99 illustrates his appliance for moving outward two first bicuspids.

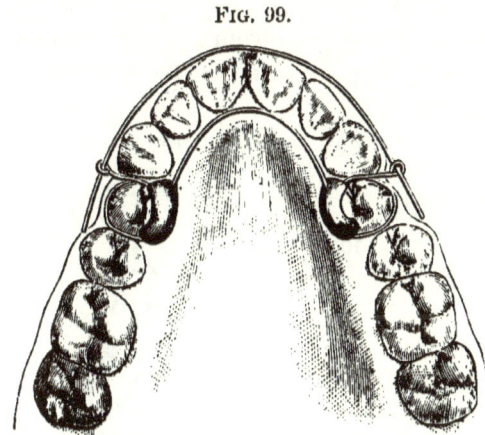

FIG. 99.

Spring and Crib Appliance. (Jackson.)

*"A base wire is shaped to the lingual side of the anterior teeth and anchored to the bicuspids by means of single crib appliances. To each of these latter is attached a hook or eyelet to sustain a straight bar of spring wire that is sprung over the anterior teeth." By this means the bicuspids may be moved outward and the arch flattened in front at the same time when desired.

* *Dental Cosmos*, Vol. XXXIII., p. 1077 et seq.

For moving the bicuspids inward, he employs a device like that shown in Fig. 100.

"Thin metal is fitted to the labial surfaces of the teeth to be moved, being made to extend well towards the necks and distal surfaces of the teeth. A good-sized spring wire is then formed to follow the outline of the anterior teeth on their labial surfaces and extend to the metal clasps, to which it is soldered.

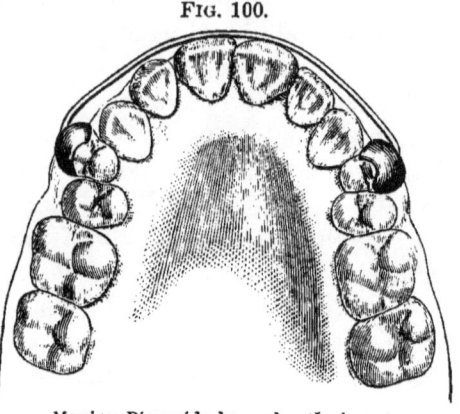

FIG. 100.

Moving Bicuspids Inward. (Jackson.)

"The appliance should be removed from time to time and the clasping ends of the spring bent toward each other to exert the pressure required."

A simple wire fixture, by the same writer, for moving either outward or inward a single bicuspid is shown in Fig. 101.

"A spring wire is bent in the form of a crib surrounding the misplaced tooth and an adjoining one on each side, passing well up toward the gum on the labial and lingual sides, with the ends of the spring wire terminating and overlapping upon the tooth to be moved.

FIG. 101.

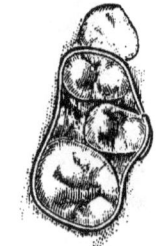

Simple Wire Spring. (Jackson.)

"The elasticity of the spring will exert the necessary force to move the tooth."

CHAPTER VI.

TORSION.

The term torsion, as applied to the teeth, signifies that condition in which a tooth is found to be turned upon its axis. Rotation refers to the act of twisting or turning a tooth so as to bring it into normal position. Torsion, therefore, describes the condition, and rotation the operation.

Torsion is usually due to some abnormal influence operative before or during eruption. Lack of space will often impel a tooth during eruption to turn in such a way as to present its smaller diameter toward the space intended for its accommodation, in order to occupy that space at all. A root, or even a portion of one, will also often cause a tooth to partly turn in its socket while seeking its position in the arch. Torsion of the superior central incisors, so often met with, is doubtless due in the majority of cases to undue thickness of the median alveolar septum. The condition is also produced after eruption by the crowding of adjoining teeth, induced by some unusual pressure, such as the effort of a later erupting tooth to occupy its place in the arch.

Torsion is met with in all degrees of extent, from the slightest prominence of one corner of a tooth to a complete half-turn.

It occurs generally in single-rooted teeth, or in those with a slightly bifurcated root; and among these, those with roots most nearly round are the ones commonly affected on account of the ease with which they can be made to turn upon their axes.

At times cases are met with in which two adjoining teeth are thus affected, usually each in like degree, this variety of the condition being known as Double Torsion.

Rotation is usually not a very difficult operation in itself, but when complicated by the crowding or disarrangement of adjoining teeth it sometimes proves quite troublesome.

Where there is sufficient space in the arch to accommodate the tooth after it has been turned, we have simply the matter of rotation to deal with; but when such is not the case, our first efforts must be directed toward providing space. This may be done, if the deficiency be slight, by pressing apart the impinging teeth by some of the means described on page 132; but where great space needs to be provided, and expansion of the arch is not indicated, it will be necessary to extract some less important tooth to afford opportunity for bringing the turned tooth into line. In the case of teeth with flat crowns, as the incisors, we may adopt either of two plans for turning the tooth, viz.: grasping the crown throughout its entire circumference and applying suitable power, or by direct pressure upon one or both of the angles that are out of line. With teeth having round crowns, such as the cuspids, we are limited to the plan of making attachment to the periphery of the crown.

At one time it was difficult, if not almost impossible, to grasp a tooth so securely as to have the attachment resist the strain of the applied power, but since the introduction of the Magill band this greatest of all difficulties associated with rotation has been overcome.

One of the simplest and most effectual methods of rotating a flat-crowned tooth is by the use of a rubber plate made to cover the palate and envelope the posterior teeth on either side, according to the Coffin plan. To the palatine portion of the plate a piano-wire is attached so as to bear upon the inner corner of the tooth to be turned, while a similar wire embedded in the buccal portion of the plate is arranged to press upon the corner that projects. The bending of the wires from time to time, to increase the tension, will speedily accomplish the desired result.

Where only one corner of a tooth stands out of line, the

plate just described may be modified by having but a single wire to press inward the outstanding corner, and allowing the rubber plate to rest firmly against the corner that is in line, to prevent it from turning.

Opportunity for the projecting portion of the tooth to move inward, must, of course, be provided by cutting away the rubber plate at this point.

Another way of rotating a tooth, is to fit a band or ferrule of gold or platinum to it, with a headed platinum tooth-pin soldered to its labial face near the angle that is out of line. A delicate vulcanite plate is then made to fit the roof of the mouth, and into it at a suitable point is screwed a threaded gold wire with a slight curve or hook on its end. After the band is cemented to the tooth, it is connected with the gold hook in the plate by means of a rubber ring. Should it be desirable to change the point of attachment on the plate, it can be done by drilling a new hole at the desired point, and screwing a hook into it. The plate can be removed for cleansing and new rubber rings applied by the patient. This plan is effective in cases where no great power is required.

FIG. 102. 
The Author's Rotating Device.

To avoid the inconvenience of wearing a plate during the school-age, the author many years ago devised a small and inconspicuous appliance for rotating a single incisor. It is shown in outline in Fig. 102, and is constructed as follows:

A strip of platinized gold about an eighth of an inch in width, and gauge No. 24 in thickness, is bent to conform to the outline that we wish the turned tooth and its neighbor to describe when in normal position. Each end of this strip is bent to partly encircle the disto-palatine angle of each tooth, after which another strip of gold, of similar width but thinner, is soldered to the centre of the first piece. This last piece should be long enough to extend between the teeth and embrace the protruding edge of the tooth to be turned.

By bending this arm so short that the appliance will have to be sprung into place, pressure is brought to bear upon the tooth that will cause it to rotate in its socket. The appliance should be removed each day, the length of the arm shortened by bending, and replaced. To guard against loss or accident, a ligature of sewing silk should be tied around the neck of one of the teeth and made fast to the appliance. About ten days will usually suffice to bring the tooth into proper position.

The tooth, once in place, is readily retained by means of the small retainer shown in Fig. 103. In its construction, similar bands are made to fit both the corrected tooth and its neighbor, after which they are joined by solder at the point where they touched when in place. To add stiffness, another strip of gold should be soldered to the palatine surface of the fixture. When completed and polished, it is lined with phosphate of zinc and placed in position upon the teeth.

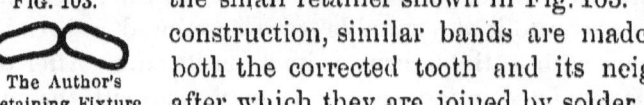

Fig. 103. The Author's Retaining Fixture.

By the use of this retainer, which occupies but little space, the tooth is held so rigid in its new position that it becomes firm much more rapidly than it would under other circumstances. Should the force exerted by the effort of the corrected tooth to return to its former malposition be so great as to affect the tooth used as anchorage, this tendency may be prevented by soldering a spur of gold to the appliance at a suitable point, and allowing this to rest against some firm tooth near by.

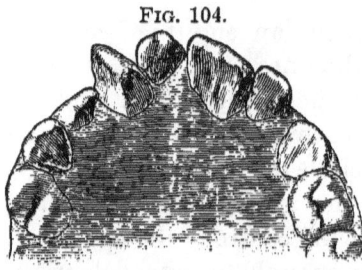

Fig. 104. Torsion Caused by Supernumerary.

A case in the practice of the author will illustrate a ready means of correcting an extreme case of torsion. The patient was a Japanese boy, nine years of age, whose upper denture when he applied for treatment presented the appearance shown in Fig. 104. The left deciduous lateral was still in place, while the right permanent lateral was just appearing through the gum. Both permanent centrals were fully erupted, but owing to the presence of a supernumerary tooth in the centre of the arch the right central was crowded far out of its place and turned upon its axis.

After extracting the supernumerary and the deciduous lateral, platinum bands were fitted to the centrals, with a gold hook soldered to each at points that would furnish the greatest amount of tractile power. After the bands were cemented in place a rubber ring was stretched from tooth to tooth, in the manner shown in Fig. 105. The malposed tooth was thus readily brought into contact with its fellow, and at the same time considerably straightened. Its further and complete rotation was then accomplished by an appliance somewhat similar to that shown in Fig. 102, after which it was retained by the retainer shown in Fig. 103. As the left central had been somewhat loosened in the act of rotating its fellow, it was found necessary, in order to secure stable anchorage, to attach a spur to the appliance and have this rest against the palatine surface of the right lateral, which was by this time almost fully erupted. In six months the teeth were firm in their new position, as shown in Fig. 106.

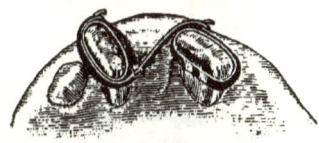

Bands and Rubber Ring for Rotation.

Corrected Case.

A simple and very effectual method of accomplishing the rotation of any tooth, without regard to the form of the crown, and one, too, in which the use of a plate is dispensed with, is illustrated in Fig. 107.

It consists of a platinum or gold band made to fit the tooth to be rotated, and having an extension bar of heavy platinized gold soldered to its labial surface. The free end of the bar is perforated by two holes for ligation to some firm tooth, usually a molar. In use, the band is cemented to the tooth and the bar sprung down and ligated to the tooth selected for anchorage. The immense leverage of this bar will quickly compel the tooth to turn in its socket. As its force becomes spent from time to time the bar can be bent outward with pliers, without removing it from the tooth. After the tooth has been brought into proper alignment, it is most conveniently held in position by means of the retainer shown in Fig. 17. It may also be retained by a rubber plate having a gold spur to pass between the teeth and rest upon the portion of the tooth that has been moved inward.

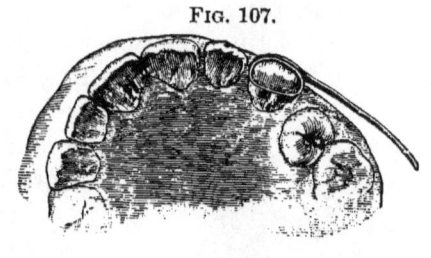

Fig. 107.

Spring Bar and Band for Rotation.

Prof. Angle has improved this appliance by making the band and bar detachable.

The band is fitted with a section of German silver tubing soldered to its labial surface, parallel with the cutting edge of the tooth. Another band, with a hook or catch soldered to its buccal surface, is fitted to a bicuspid or molar. This latter band also has a piece of tubing, soldered horizontally to its palatine surface, through which is passed a piece of wire intended to rest against the two teeth adjacent to the one banded and thus afford greater resistance. After both of these bands are cemented to their respective teeth, a straight piece of piano-wire is inserted in the tube of the tooth to be turned, and bent down and caught in the catch on the other

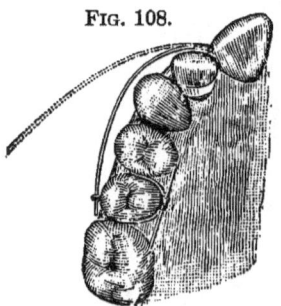

Fig. 108.

Rotation. (Angle.)

tooth, as shown in Fig. 108. The advantage of this modification is, that a weaker or stronger wire can be substituted at will, and the power be thus readily controlled. When the tooth is in proper line, the wire is removed and replaced by a shorter one resting upon an adjoining tooth. This acts as a retainer by keeping the tooth in position until it has grown firm. The retaining wire is secured by means of a pin, inserted in a hole drilled through both tube and wire.

Another simple device devised by Dr. Jackson * for rotating a single incisor is illustrated in Fig. 109. It consists of a band or collar, made to encircle the one incisor, to which are attached upon the labial and palatal surfaces two lugs to receive a U-shaped wire. One arm of this wire spring lies upon the labial side of the teeth and produces pressure upon the mesio-labial corner of the turned tooth while the other extends along the palatal surface and presses upon the disto-palatal angle. Pressure in opposite directions is thus accomplished, while the balancing of the two forces prevents the anchor tooth from turning. The appliance is too small to in any way interfere with speech or occlusion.

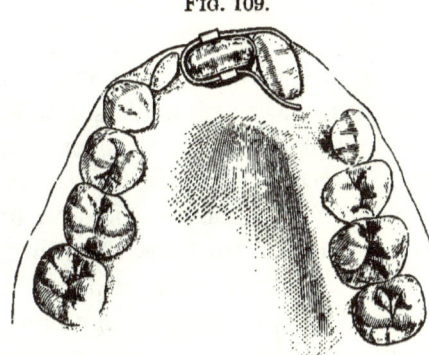

Fig. 109.

Rotating Device. (Jackson.)

DOUBLE TORSION.

Where two adjoining teeth, as the superior centrals, are to be rotated in opposite directions, a single appliance will often accomplish both movements at the same time. The appliance devised by the author for this purpose is shown

* *Dental Cosmos*, Vol. XXXIII., p. 1076.

in Fig. 110, and the details of construction in Fig. 111. It is a modification of the appliance for single rotation shown on p. 159. To adapt it for duty in turning two teeth, instead of the single strip of gold passing between the teeth, two strips are bent in the form of "b" and "c." These are made long enough to be bent slightly over the labial surfaces of the teeth to be turned, extend along the mesial surface to the palatine, and then along this latter almost to the distal angle. After being properly shaped according to the model, they are clamped together and soldered along their contiguous surfaces.

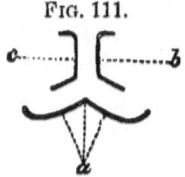

Fig. 110.
The Author's Device for Double Rotation.

Fig. 111.

This part is then placed in position on the model, and the long arms bent to conform to the inner surface of the bar "a," after which it is removed, soldered to "a," and the part "b" "c" reduced in thickness by filing, so as to occupy as little space between the teeth as possible. When properly constructed the labial part of the appliance will rest against the teeth just at or slightly above the most prominent points of their convexity, while the lingual portion will be near the gum, but not quite touching it, and the slightly curved ends of this part will catch just above the little prominence usually found at the disto-palatine angle near the gum.

Thus made and placed, the piece cannot become displaced by the lip or tongue, except when it has become loosened by the moving of the teeth. As will readily be seen, by its use force is brought to bear upon four points of the two teeth at one time, acting as a double lever upon each tooth.

A valuable feature of the appliance, had in view in its devising, is that it occupies but one interdental space, and thus more readily favors the turning of teeth that are more or less crowded.

In use, the patient should be seen each day, the fixture

removed and tightened by bending the long arms slightly toward the smaller ones and sprung into place.

To facilitate its introduction in the first instance, a piece of rubber should be placed between the teeth one day previous to the insertion of the appliance.

As in the case of the appliance for single rotation, a thread should be tied around one of the teeth and attached to the front bar to guard against the swallowing or loss of the piece. Fig. 112 represents a case of double torsion which was corrected in ten days' time by the use of the appliance just described, the patient being seen every day; while Fig. 113 shows the completed operation. After the teeth are in posi-

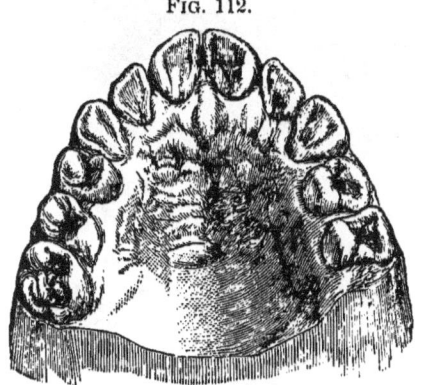

FIG. 112.

Double Rotation.

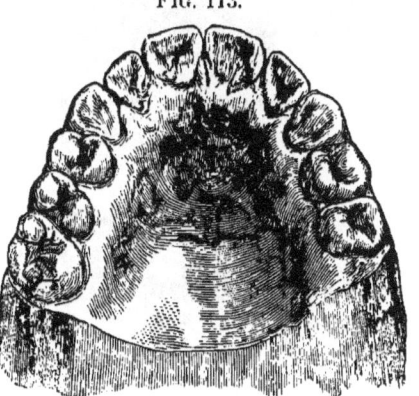

FIG. 113.

Corrected Case.

tion, they may be best retained by means of the retainer shown on p. 160.

When the distal corners of the teeth project instead of the mesial, the appliance described is rendered equally serviceable by reversing its position and placing the long arm on the labial surface. Fig. 114 represents a case of this character, while Fig. 115 shows the rubber plate with gold wire bow that was used to retain the teeth after correction. A simpler and better method of retention would have been to use the appliance shown in Fig. 103.

Prof. Angle has devised a very simple and effectual method of accomplishing double rotation where the mesial angles protrude. Upon each of the teeth to be rotated he places Magill bands with tubes soldered to their labial faces near the distal angles. One tube is set vertically and the

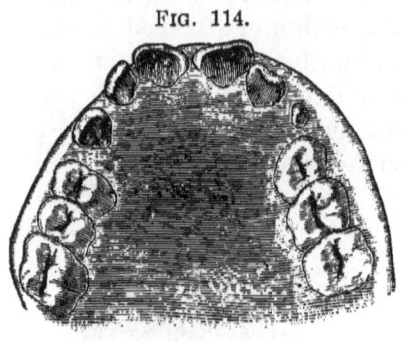

Fig. 114.

Torsion of Centrals, with Distal Angles Pointing Outward.

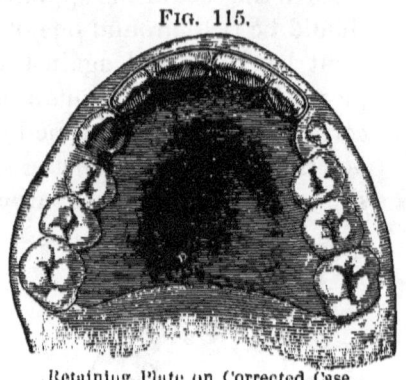

Fig. 115.

Retaining Plate on Corrected Case.

Fig. 116.

Fig. 117.

Angle's Appliance for Double Rotation.

other horizontally. A short piece of piano or German silver wire, bent to a right angle at one end, is inserted into these tubes and rotation is effected by the elasticity of the wire.

Two views of the appliance are shown in Figs. 116 and 117.

Once in position, the teeth are retained by inserting in the tubes a suitably-shaped piece of non-elastic gold wire.

CHAPTER VII.

CONTRACTION OF THE ARCH.

A contracted arch may be due to lack of development, caused by late or mal-eruption of some of the teeth; to the loss of certain of the permanent teeth soon after their eruption; or to malposition of the teeth in the opposite jaw.

The late eruption of the superior cuspid teeth, where their spaces have been preöccupied by teeth anterior and posterior to them, is perhaps the most frequent cause of this deformity.

In some cases, the contraction is limited to the molar and bicuspid region; in others, to the anterior alone; while in others still, the entire arch needs expansion.

The enlargement of the arch, either at certain points or in its entirety, may be accomplished by a variety of methods.

Where lateral expansion is desired, it may usually be brought about in a simple manner by the use of the Coffin split-plate, the construction and operation of which are described on p. 108.

Another form of appliance, intended to accomplish the same purpose and constructed of piano-wire and vulcanite, has been devised by Dr. Talbot, and is illustrated in Figs. 118 and 119.

In his description, he says:* " A (vulcanite) plate is made to fit the teeth and alveolar process, and cut away so that the anterior parts extend far enough forward to enclose the teeth to be moved. A piece of (piano) wire is bent into either of the forms shown in Fig. 119, wherein ' a ' is the coil and fixed point, ' b b ' movable arms extending from ' a,' and ' c c ' movable arms extending from ' b b.' Grooves

---
* Talbot's Irregularities of the Teeth. P. 221, 2d Ed.

are cut into the anterior and posterior parts of the plate to correspond with and receive the points 'b b' and 'c c.'

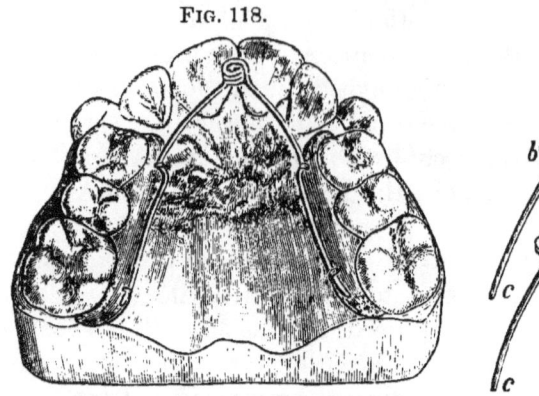

Fig. 118.

Talbot's Appliance for Lateral Expansion.

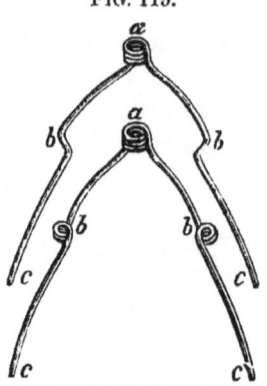

Fig. 119.

Talbot Springs.

Holes are drilled at these points, and the wires tied to the rubber plates. In order that the anterior teeth may be moved with the greatest force, the arms are so adjusted that the greatest pressure is exerted on the anterior parts of the plates. This appliance is readily removed for cleansing, and returned to place by the patient."

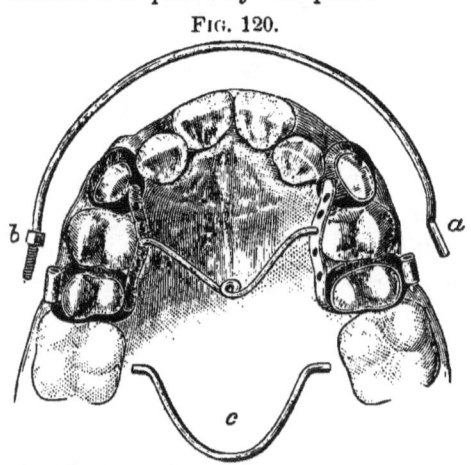

Fig. 120.

Combination Appliance for Expansion. (Goddard.)

Prof. Goddard employs the Talbot spring for lateral expansion, dispensing with the rubber plate and using instead, band and bar attachments to the teeth, as shown in Fig. 120. The cut so fully illustrates the appliance that very little explanation is needed.

The holes in the bars on either side are for the reception of the coiled spring which can be placed either forward or backward, according as one part calls for more expansion than the other. After the arch has been widened, the bent wire "c" is substituted for the coiled spring, and retains the advantage gained. The long wire "a" "b" is intended to be used where any of the incisor teeth need to be moved forward. In such case its ends are inserted into the tubes on the bicuspid bands and rubber rings are passed over it and the in-lying incisors. To prevent the ends of this wire from slipping through the tubes they may be threaded and supplied with a nut, as shown at "b," or they may be bent to a bayonet-shape, as shown at "a."

A very similar appliance for use in the lower arch, also devised by Prof. Goddard, is illustrated in Fig. 121.

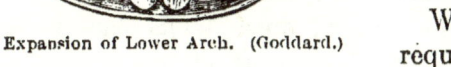

Fig. 121.

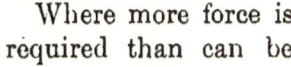

Expansion of Lower Arch. (Goddard.)

In this case, as in the other, the long wire was used for attachment in drawing forward the in-locked laterals.

Where more force is required than can be obtained from either of the appliances just described, it can be had by the more direct power of the jack-screw, operating upon the portions of a rubber plate lying next to the teeth to be moved.

Dr. Kingsley's neat and effective appliances of this character are shown on p. 154.

The use of the jack-screw in the lower jaw would appear to be objectionable on account of its being in the way of the tongue, but experience has proven that this objection is, in fact, a slight one.

The use of the screw hastens the operation and thus lessens the period of inconvenience in any given case.

When expansion of the anterior portion of the arch is

desired, it may be accomplished by means of one of the appliances shown on pp. 134 and 135, or by a modification of the Coffin split plate devised by Prof. Goddard. The latter is shown in Fig. 122.

As will be seen, there are two corrugated piano-wires attached to the rubber plate, one on each side near the free margins, while the plate is split laterally just back of the incisor teeth. As in other split plates for anterior expansion, this plate is made in one piece and the wires arranged so that their anterior ends are embedded in the portion to be detached, while the posterior ends are fastened to the main body of the plate.

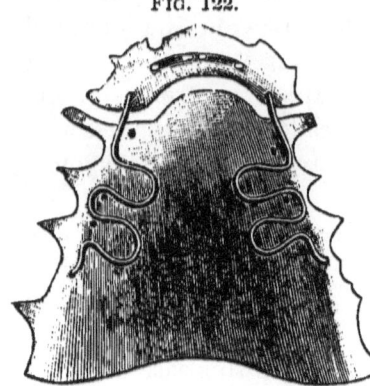

Fig. 122.

Goddard's Split Vulcanite Plate.

After the completion of the plate the front portion is separated by means of a jeweler's saw, and pressure is produced by stretching the wires from time to time.

The anterior portion is kept down to its place by being ligated to the central incisors. In using this form of plate the author has found it more convenient to hold the front portion down by imbedding in the plate a gold spur, to pass between the centrals in the free space near the gum. He also prefers to secure the main portion in position by making the plate to cover and grip the bicuspids and molars, as in the Coffin method, instead of fastening it to the side teeth with ligatures. The appliance is admirably adapted to the purpose for which it was devised.

Where expansion of the entire arch is desired, a better plan, in most cases, is to accomplish it by separate operations. Lateral expansion, for instance, can be accomplished first, and after the bicuspids and molars have been brought into proper position, they may be retained by means of a rubber plate covering them. This plate will not only hold them firmly, but serve as an anchorage to which other fixtures may be attached for the expansion of the anterior portion of the arch, as in the Goddard plan.

The details of a case of general expansion of the superior arch, may be of interest to the student.

The patient was a boy of about fifteen years of age. The inferior arch was of normal size, with the teeth well arranged. In the superior arch, all of the teeth except the cuspids articulated inside of the lower ones, giving the patient a pinched or contracted appearance in the region of the upper lip. The laterals were almost in contact with the first bicuspids, while the cuspids had fully erupted outside of the arch and were overlying the laterals.

Extraction was not indicated, for all of the teeth were needed to fill the arch after its expansion.

By means of a Coffin split-plate, lateral expansion was accomplished in about a month, so that the bicuspids and first molar on each side occluded normally with those below. Next, with another Coffin solid plate encasing the teeth that had been moved, and with two piano-wires attached, the laterals were pressed forward: after which, new rubber was added to the plate to keep these teeth in position, and the wires changed to press the centrals forward into line with the laterals. After this had been accomplished there was still insufficient space for the accommodation of the cuspids, and as the incisors were already so far forward that pressure could not advantageously be brought to bear upon them from the rear, another plan for increasing the cuspid space was decided upon. Magill bands were made to fit the laterals, with gold spurs extending along the palatine surfaces of the centrals to insure uniform movement of the four incisors. Platinum bands were also attached to the first bicuspids. All of these bands we reënforced with an additional piece of platinum soldered to the portion next to the space. Through these re"nforcements, at about the centre of the tooth, holes were drilled entirely through the bands. Piano-wire was next bent into the form of small U-shaped springs with the ends at right angles, similar to Dr. Talbot's plan, but without the coil. Grasping these near the neck with a pair of narrow-beaked right-angle forceps transversely grooved near the points to seize the wire, the springs were placed in position with their ends resting in the holes of the bands. As, from time to time, the force of these springs became spent, they were removed and their power renewed by enlarging their curves. Sufficient additional space having been gained by their use, the cuspids were forced into position by means of a Coffin plate with wires attached to the buccal surfaces, extending forward and resting upon the labial surfaces of the cuspids.

The appearance of the arch and teeth with the U-springs in position, is shown in Fig. 123. The operations were not hurried, and consumed about one year's time.

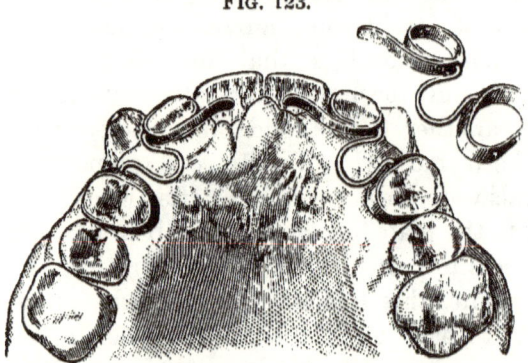

Fig. 123.

Increasing Space by Curved Spring and Bands.

A retaining plate of vulcanite covering the roof of the mouth, with gold loops attached to overlie and retain the cuspids, was worn for nearly a year.

Another case, differing somewhat from the one just given, was that of a young girl about eleven years of age, whose superior arch did not need lateral expansion, but required anterior enlargement to accommodate the in-coming cuspids. False occlusion of the superior incisors also needed correction. Fig. 124 represents the case as it presented. The superior centrals met the lower ones edge to edge, while the superior laterals passed inside of the lower ones. There was very little room between the superior laterals and first bicuspids to accommodate the cuspids, which, slow of eruption, were just beginning to make their appearance.

Fig. 124.

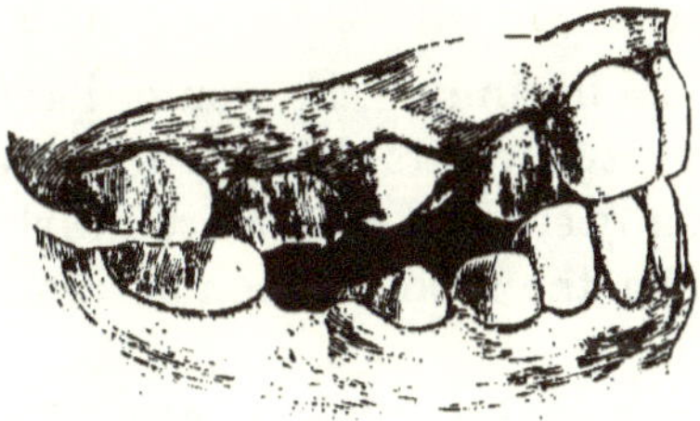

Case Requiring Anterior Expansion.

The treatment required was the moving of the laterals and centrals so as to overlap the lower ones, and the moving backward of the bicuspids on each side to afford space for the cuspids. The laterals were first moved forward into line with the centrals, by means of the plate shown in Fig. 125. This accomplished, the anterior portion of the arch was expanded by means of a Goddard split-plate.

A plain rubber plate, covering the arch and touching each tooth, was next made, and into it were secured on either side pieces of piano-wire bent to right angles at their free ends, the bent portions being arranged to rest upon and press against the mesial surfaces of the first bicuspids to force them backward. The plate having been trimmed to admit of the backward movement of the bicuspids, full space for the cuspids was soon gained.

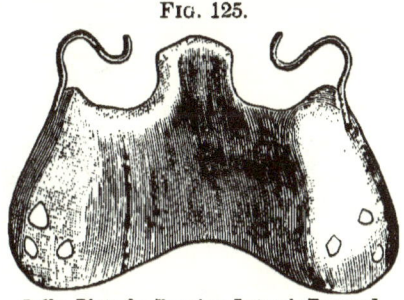

Fig. 125.

Coffin Plate for Pressing Laterals Forward.

The slow eruption of the cuspids required a retaining plate to be made, armed with gold spurs at suitable points, to keep the regulated teeth in their new positions and await the full eruption of the cuspids.

At the end of six months the cuspids had assumed, unaided, their proper places in the arch, and by their key-like position preserved the arrangement without the further aid of any retentive appliance.

The case had previously been in the hands of two dentists, who began operations for correction, and it therefore became necessary for the author to carry it forward to completion.

Had he been consulted in the beginning, he would have advised non-interference until two years later when the cuspids would have been partially erupted, and more nearly ready to assume their places in the arch, as soon as room was provided.

In this way the wearing of a retaining plate, to await the full eruption of the cuspids, would have been avoided and the case simplified.

Prof. Angle has devised a neat and effective appliance, constructed entirely of metal, for the lateral expansion of the arch, as shown in Fig. 126.

Like the Jackson appliances, it utilizes the principle of the Coffin spring without the objectionable features of the rubber plate. It can be used in either the upper or lower arch and where no greater power than the spring affords is needed, will prove very efficient.

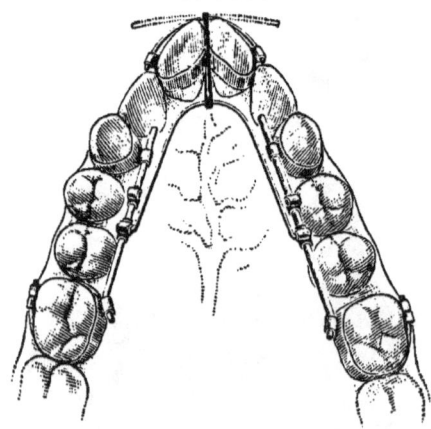

Fig. 126.

Lateral Expansion. (Angle.)

As seen in the cut, a rubber ligature may be attached to the centre of the spring and be connected with any cross-bar appliance upon the incisors for drawing them inward when such additional movement is desired.

CHAPTER VIII.

PROTRUSION OF THE UPPER JAW.

This deformity, so frequently met with in our day, not only destroys all harmony of expression, but so strongly suggests the facial characteristics of idiocy as to be particularly objectionable.

Fig. 127 shows the relation of the teeth in outline and Fig. 128 the facial expression. In the latter will be noticed the conspicuousness of the superior incisors and the resultant shortening of the upper lip.

FIG. 127.   FIG. 128.

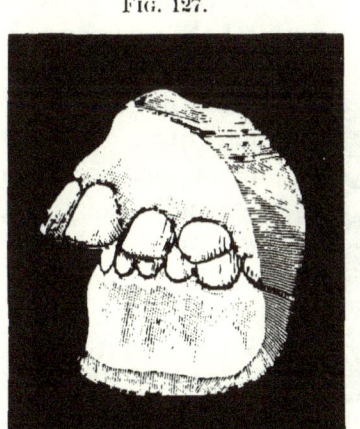 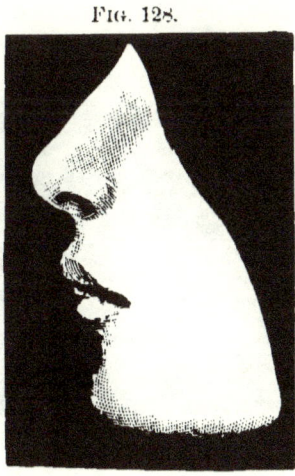

Superior Protrusion. (Case.)

The causes tending to produce this condition, have been briefly considered on pages 23 and 24.

There are two varieties of this deformity:—

1st. Where the lower teeth are in line forming the normal curve, while the upper ones pass over and beyond them so as not only to interfere with enunciation, but also to render them almost unserviceable in mastication. This form is usually attributable to inheritance; to the abnormal size of the teeth in the superior arch; or to the mechanical influence of pressure on the part of the posterior teeth. It is the one most easily corrected, on account of the operations being confined to a single arch.

2nd. Where the lower incisors are flattened in outline or introverted, and the superior ones extend so far forward as to leave a large space between the two when the jaws are closed. In this case, the superior protrusion appears to be greater than it really is, on account of the superior and inferior teeth inclining in different directions. Where there is introversion of the inferior incisors we generally find their cutting edges on a higher plane than that of the neighboring teeth.

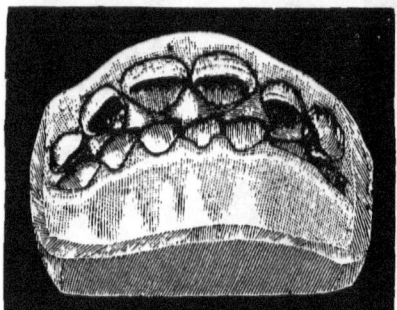

Fig. 129.
Deep Underbite. (Case.)

This condition is, in most cases, due to the habit of thumb-sucking, the thumb pressing the lower ones in and the upper ones out at the same time.

The relatively greater height of the cutting edges of the lower incisors causes them, in most cases, either to occlude with the bases of the crowns of the superior incisors, or to come in contact with the soft tissues back of them, as shown in Fig. 129. This condition seriously complicates the matter of correction, for it interposes an obstacle to the inward movement of the superior teeth and the outward movement of the inferior ones.

Where the protrusion is slight and the teeth are in contact, space for their inward movement may sometimes be obtained by dressing off any discoloration or superficial decay from the approximal surfaces of the six anterior teeth with sand-paper discs or emory-cloth strips, followed by thorough polishing.

By this means the author has, in a few instances, materially improved the patient's expression, without loss of teeth or injury to tooth-substance. The space once gained, the teeth can easily be brought inward by the use of a Coffin plate, cut away posteriorly to the incisors, and having gold hooks attached to the anterior portions of the plate on the buccal surface. A rubber band caught over the hook on one side, carried along the labial surfaces of the anterior teeth and attached to the hook on the opposite side, will generally provide the required tension. Small double hooks, made from half-round gold wire and hung over the cutting edges of the centrals, will, by their second curves, support the rubber band in proper place and keep it from resting upon and irritating the soft tissues. Other simple means for effecting the same result, will readily suggest themselves to the operator. Where the protrusion is of greater extent and the teeth are in contact, it will be necessary in most cases to sacrifice a bicuspid or molar on one or both sides of the mouth to obtain sufficient space to enable the anterior teeth to be moved backward into line.

After the extraction of the tooth or teeth it is well to draw backward, by easy stages, the teeth on either side anterior to the space, to and including the cuspids. The subsequent drawing in of the four incisors will then be a comparatively easy matter. In many cases, if the posterior teeth were used as anchorages for the inward movement of six or ten anterior teeth, they would be more likely to move forward than to cause the anterior ones to be forced backward, on account of the disparity of resistance.

A number of methods for moving backward the cuspid and bicuspid teeth are described on pp. 141 and 143. A simple plan for drawing in the four superior incisors, is shown in Fig. 130. Platinum bands are fitted to the laterals, and to their labial portions are soldered extensions of gold, to cover and rest upon the labial surfaces of the adjoining centrals.

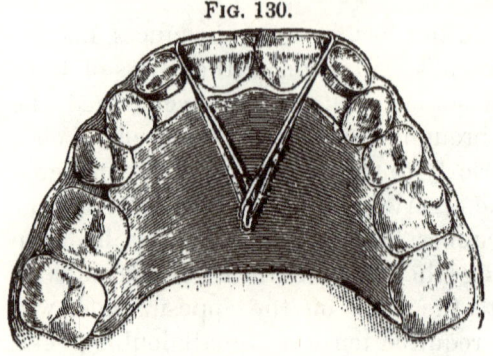
Fig. 130.
Author's Device for Retracting the Superior Incisors.

A plain rubber plate is also made to cover the palatine arch, with a gold hook inserted in its centre. The bands being cemented in place, rubber rings are slipped under the extensions and carried to a point between the centrals and laterals, where they are drawn in and over the gold hook in the plate. By their contraction, all four of the incisors are drawn inward while but two of them are banded.

A plan differing somewhat from the one just described, is that of Dr. Kingsley's, illustrated in Fig. 131. The band overlying the incisors is of gold, and has hooks soldered to the upper edge to prevent its slipping up to the gum. It is also fitted with a thin strip of gold to pass between the centrals, the free end of which is connected with the centre of a vulcanite plate by means of a ring cut from rubber tubing. This rubber ring is made fast to the plate either by a ligature or by slipping it into a horse-shoe slot cut in the plate for the purpose.

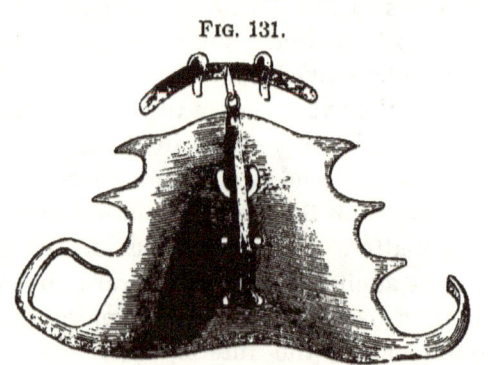
Fig. 131.
Kingsley's Gold Bar and Vulcanite Plate for Retraction.

In many cases the elasticity of rubber does not provide sufficient force to move backward the anterior teeth. In such event advantage may be taken of the superior power furnished by piano-wire. An excellent plan for arranging and anchoring such wires is furnished by Dr. Wadsworth.* The appliance is constructed after the method of Dr. Jackson, and is illustrated by Fig. 132.

In his description, Dr. W. says:—

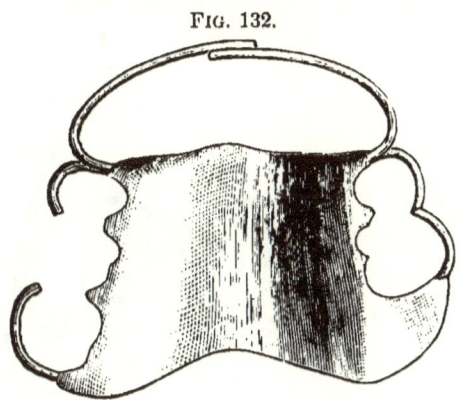

Fig. 132.

"A vulcanite plate was fitted to the roof of the mouth, and well cut away from the lingual surfaces of the front teeth. Piano-wire springs of No. 21 wire were vulcanized into the plate, passing through the spaces made by the removal of the first bicuspids, and following around from each side of the labial surfaces of the cuspids and incisors were made to pass each other at the median line. These springs were bent to give the required pressure on the teeth to be moved and the pressure increased by bending the springs from time to time as the teeth were moved inward. The appliance was held in position by clasps made from No. 20 piano-wire fitting the bicuspids and molars, as seen in cut.

"The patient was seen once or twice each week to increase the pressure as required and the deformity entirely corrected in three months and a half. The appliance was worn continuously, could easily be removed for cleansing and replaced by the patient, and caused no pain or inconvenience whatever."

---

* *Dental Cosmos*, Vol. XXXIII., p. 30.

The direct and forcible action of the screw may be conveniently brought into play by means of the device shown in Fig. 133. It is a vulcanite plate covering the arch and encasing the molars, to which is attached a half-round gold wire bent to a curve and long enough to extend along the outer surfaces of the teeth from molar to molar. One end of this curved wire is permanently attached to the vulcanite plate while the other terminates in a threaded wire, which engages with a gold nut playing in a slotted recess of the plate on the opposite side. Turning the nut shortens the bar and draws the teeth inward.

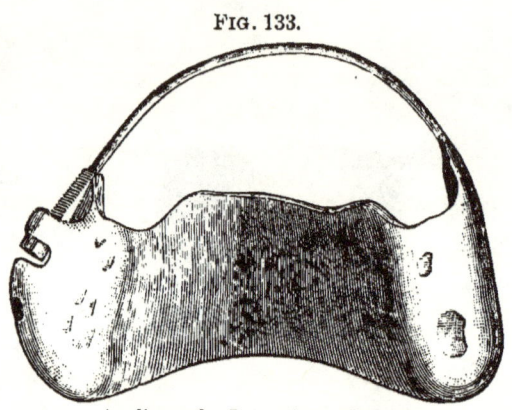

Appliance for Retraction. (Tomes.)

When still greater power is demanded, as in cases where it is desired to draw the six anterior teeth inward by one operation, or where the incisors do not yield readily to any power that can be applied within the mouth, anchorage for resistance must be obtained outside. Dr. Kingsley, we believe, was the first to suggest and utilize the back of the head as an anchorage for appliances intended to produce movements of the teeth. Illustrations of a fixture of this character will be found in his work, pp. 133 and 134.

Dr. Farrar also devised an apparatus for the same purpose, but it is somewhat complicated in its construction and manner of adjustment.

One of the simplest devices of this character, is that of Prof. C. L. Goddard.* In describing the construction and

---

* Annual of the Universal Medical Sciences, for 1888, Vol. III., pp. 547–551. F. A. Davis, Philadelphia, publisher.

use of his appliance, he says: "On a cast of the superior incisors a small sheet of wax was placed, covering the labial surfaces, cutting edges and part of the lingual surfaces. In the anterior surface of this wax plate, a steel wire was embedded, curved to conform to the arch, and extending laterally about one inch and a half on each side. The ends of this wire were bent in the form of hooks. The wax plate and wire were then embedded in a flask by bending the ends of the wire sufficiently to allow them accommodation inside of the flask. By the methods usually employed in vulcanite work, a plate was thus made of black rubber with the wire attached, as shown in Fig. 134.

Fig. 134.

Goddard's Steel and Vulcanite Appliance for Retraction.

"When placed on the patient's teeth, the ends of the wires projected from the corners of the mouth on each side far enough to permit elastic bands to connect them with a cloth cap on the patient's head without touching the cheeks.

"The cap was so shaped that the elastic could be attached to it in two places on each side, one above and one below the ear, by means of dress hooks sewed to the cap at these points. Round silk-covered elastic cord was used, and the direction of the force could be varied by using a greater number of strands above or below the ear, according to the requirements of the case. The amount of force was easily varied by shortening or lengthening these cords. Fig. 135 shows the appliance in position.

"This appliance was worn at night only, and the teeth were soon moved back to the desired position. The inferior incisors striking the bases of the superior ones, were moved backward with them. After the teeth were in proper position, the tension of the elastic cord was slightly lessened and the appliance worn at nights for a few months as a retaining appliance, until the teeth became firm.

"The greatest usefulness of this appliance is in cases where there are no teeth in the mouth sufficiently firm for the anchorage of an appliance of ordinary form, or where the teeth, if firm enough, are of such shape that it is practically impossible to fasten appliances to them."

Fig. 135.

Goddard Appliance in Position.

The worst case of superior protrusion the author ever met with was corrected by the use of an appliance differing from Prof. Goddard's only in certain minor particulars.

The patient was a boy, sixteen years of age, whose superior teeth projected beyond the lower ones at least three-quarters of an inch. The inferior incisors were relatively long, and their cutting edges, in occlusion, embedded themselves in the soft tissues of the palate quite a distance inside of the superior teeth. Both arches were wide and well-formed, and all of the teeth were in contact. Fig. 136 shows the relation between the upper and lower teeth at the time of presentation for treatment. All of the teeth being equally good the first bicuspids were removed to provide space. An appliance of vulcanite and wire, similar to Prof. Goddard's, was then made; the wire, after it was properly fashioned, being nickel-plated before vulcanization. The scull-cap, instead of cloth, was made in skeleton form of inch-wide black silk ribbon, each strip being double and lightly stuffed with cotton to make it more comfortable for the patient. The elastics used were the ordinary flat and wide rubber bands, cut and perforated near the ends to engage with the hooks on the cap. The teeth being large, strong and firmly set, especially the cuspids, moved slowly; but in five months' time, by the use of the above appliance alone, the teeth were moved back into proper position, the cuspids coming into close contact with the second bicuspids. The cutting edges of the lower incisors were ground off somewhat to enable the superior ones to be moved inward.

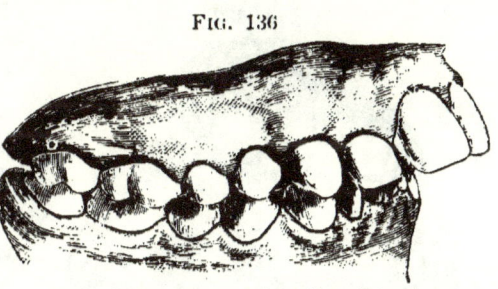

Fig. 136

Superior Protrusion caused by Thumb-Sucking.

The boy being in attendance upon school at the time of the operation, and not wishing to subject himself to the ridicule of his schoolmates that the wearing of such a conspicuous appliance would surely bring upon him, an accessory appliance was devised for him to wear during school hours. It consisted of a thin silver saddle covering the protruding centrals, to which, on the labial surface near the terminations, were soldered two platinum headed pins. The first molars were fitted with platinum bands, to which platinum hooks were also attached on the buccal surface. The bands were cemented to their respective teeth, while the saddle was removable. This appliance, in position, is shown in Fig. 137.

In use, the saddle was placed in position and the pins upon it and the molar bands connected by means of thin rings cut from French rubber tubing of small diameter. This fixture was simply intended to retain, during the day, the progress made by the more powerful appliance at night. It was put on in the morning before starting for school, and after school hours was replaced by the pressure appliance, which was worn until morning.

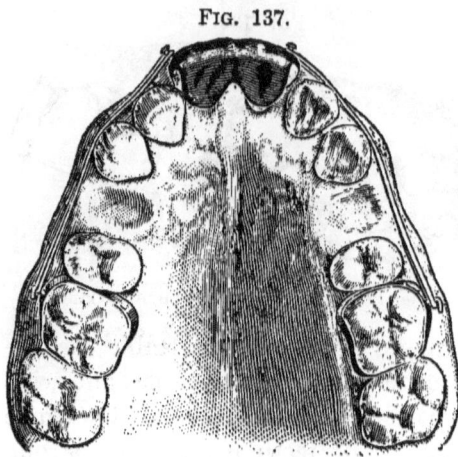

FIG. 137.

Day Retaining Appliance

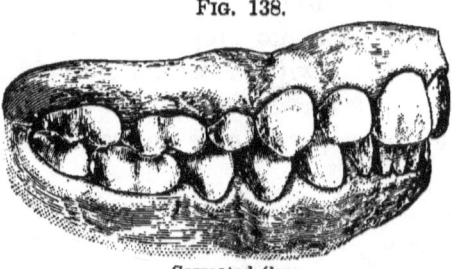

FIG. 138.

Corrected Case.

Both appliances were removable for cleansing, and were readjusted and operated by the patient himself. They gave him no pain or inconvenience to speak of, and required very little oversight on the part of the operator. For the first three months of retention the day appliance was worn both day and night, and for the succeeding three months at night only. Fig. 138 shows the relation of the teeth after the operation was completed.

For the reduction of anterior protrusion, Prof. Angle's appliance as shown in Fig. 139, commends itself for simplicity and efficiency. It consists of anchor bands (D) for the molar teeth with long tubes soldered to their buccal surfaces to receive the wire bow-spring (C) which rests in front in notched projections upon bands (A) cemented to the central incisors. At the centre of the bow-spring is soldered a short tube, having upon its labial surface a rounded projection to receive the standard (cupped at its free end) of the long traction bar (E). In use, the clamp-bands (D) are attached to the anchor teeth and the plain bands (A) cemented to the central incisors. The bow-spring (C) is now placed in position.

Angle's Appliance for Reduction of Anterior Protrusion.

Occipital resistance is obtained by means of a netted cap fastened to a circle of wire fitted to the head, to which are attached rubber bands. When the cupped standard of the traction bar has been placed over the central spur of the bow-spring, the rubber bands of the cap are drawn forward and looped over the curved ends of the traction bar, as shown in Fig. 140. This cap, traction bar and rubber bands are worn only at night on account of their conspicuousness.

During the day, rubber rings (B) are caught over the

tubes on the molar bands and secured by ligature to projections on the bow-spring in the region of the cuspid teeth. The appliance in position, as worn during the day, is illustrated by Fig. 141. After reduction of anterior protrusion we are met with the difficulty of retaining the results gained. Although the posterior teeth in many cases will not furnish the resistance necessary for drawing the anterior teeth inward, they will usually answer perfectly for retaining them afterward. Attachment can be made to them either by means of a rubber plate covering the roof of the mouth and extending around their distal surfaces in the form of a clasp, or by means of metal bands cemented to them. In the former case a small round

Fig. 140.
Night Appliance. (Angle.)

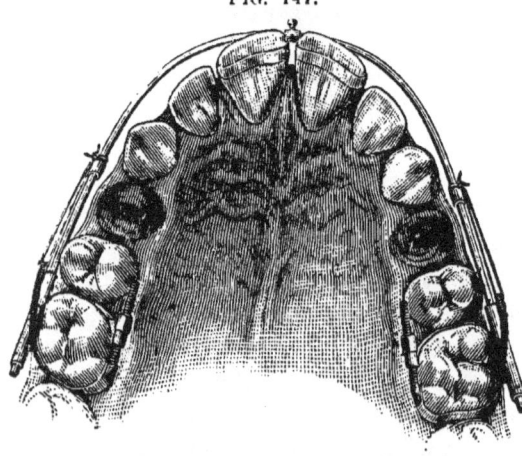

Fig. 141.
Day Appliance. (Angle.)

or half round gold wire may be made to pass around the arch, touching the regulated teeth on their labial surfaces, and be attached at each end to the rubber plate at convenient points, as where teeth have been extracted. In the latter case a similar retaining wire may be soldered to the molar bands, or the bands may have tubes soldered to their buccal surfaces and the wire, threaded at the extremities, passed through these and retained by means of nuts operating upon them. In either case the retaining wire should have short gold clips attached to it in front to engage with the cutting edges of at least two of the incisor teeth.

Where it is desired to avoid having a retaining wire pass entirely around the front of the arch, a rubber retaining plate may be made with a gold T passing between the centrals and long enough to rest upon all four of the incisors. Holding these teeth firmly in place will also keep the cuspids in line through lateral pressure.

In all cases the retaining appliance should be worn for a year or more, until we are fully satisfied that the teeth are firm in their new positions and manifest no tendency to change.

CHAPTER IX.

PROTRUSION OF THE LOWER JAW, OR PROGNATHISM.

This condition, one of the most unsightly of dental deformities, giving to the individual a rather inhuman expression and interfering greatly with speech and mastication, is quite frequently met with. The causes probably responsible for its inducement are given on p. 25.

When the deformity is slight it may be corrected, or at least modified, by pressing the lower incisors inward and the upper ones outward; but where the case is pronounced, there seems to be no remedy for it but the retraction of the entire inferior maxilla. This may best be accomplished by using some form of scull cup, and connecting it with a padded chin piece by means of strong rubber bands. The persistent contraction of the rubber will, in a greater or less time, dependent largely upon the extent of the deformity and the age of the patient, bring about the desired change.

In the accomplishment of this retraction, it was formerly supposed to be brought about by a change effected at the angle of the jaw; but the more plausible hypothesis, is the one first advanced by Dr. Geo. S. Allen, namely: That the pressure applied to the mental region causes resorption of the posterior wall of the glenoid cavity, thus permitting the condyles to recede and articulate somewhat posteriorly to their former position. This theory as to the physiological change brought about, is supported by the fact that an alteration of form in the gleniod cavity is more readily accomplished by resorption, than a bending of the maxilla at its strongest point.

An interesting case of retraction of the lower jaw, was brought before the Odontological Society of New York, in 1878, by Dr. Allen. I quote important points from his description: "As will be seen from the photograph (Fig. 142), taken at the time she was wearing this apparatus, it consists of two parts. For the lower part, I made a brass plate to fit the chin, having arms with hooked ends reaching to a point just below the point of the chin. These arms were arranged in such a way, that the distance between them could be altered at will, by simply pressing them apart or together. The upper part consisted of a simple network,

Allen's Device for Retraction of Lower Jaw.

going over the head and having two hooks on each side, one hook being above and the other below the ear. When this apparatus was completed and in use, there were four ligatures of ordinary elastic rubber pulling in such a way as to force the lower jaw almost directly backward. The work proceeded very rapidly, so that at the end of two months the irregularity was almost entirely cured. I see no reason why, in all such cases, either this or similar methods of procedure should not be adpoted. I should certainly, if any similar cases presented hereafter, even at twelve or thirteen years of age, before attempting any other procedure, try this first and thoroughly."

In forming the chin piece for cases of this character the author is accustomed to take a plaster impression of the chin and from this make a model. The model is then overlaid with a piece of trial-plate wax, from which, after being varnished, a mould in sand is obtained and a die and counterdie made. Between these a piece of soft and heavy brass plate is struck up and drilled full of holes. After fashioning heavy piano wires to cross the plate and extend sufficiently beyond to form hooks, they are soft-soldered to the brass plate and the latter covered with black silk with a thick layer of cotton batting laid between the two. The enlarged size of the chin piece will admit of this. The piece thus padded will fit the chin and be soft enough to prevent pain when pressure is brought to bear upon it.

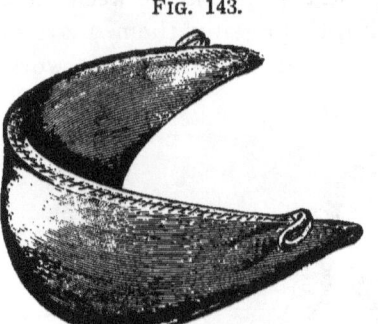

FIG. 143.

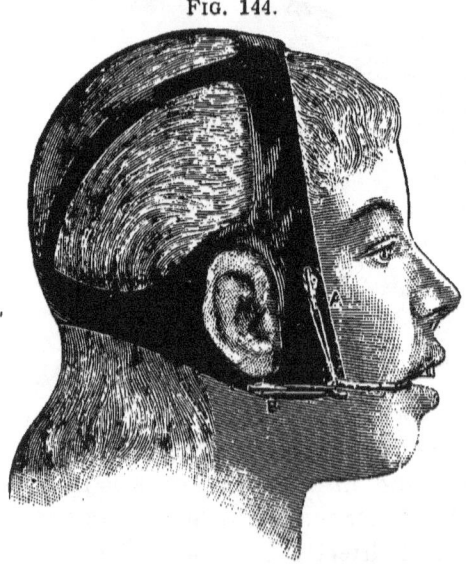

FIG. 144.

A chin-piece devised and used by Dr. Kingsley* is shown in Fig. 143. It is made of "sheet copper (stiffened around the edge with non-elastic steel wire accurately fitted to a plaster cast of the chin) padded, and covered with leather."

The skeleton skull-cap, used in connection with the chin appliance by Dr. K. is made of leather and is shown in position in Fig. 144.

The Drs. Winner, of Wilmington. Del., have furnished the writer with models and description of a case somewhat similar to the foregoing (Figs. 145 and 146). In their case, the patient was a boy fourteen years of age, tall, slender, possessing good general health, but only fair physical strength. The models show that there was a bicuspid lacking on each side above, while below there still remained two temporary molars. He stated that he had never had any teeth extracted by a dentist, so it is probable that the two bicuspids were never erupted. The superior centrals were considerably worn away on their cutting edges and labial surfaces by attrition with the lower ones. After extracting

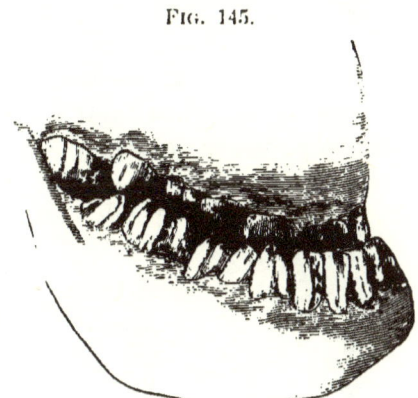

FIG. 145.

Prognathism.

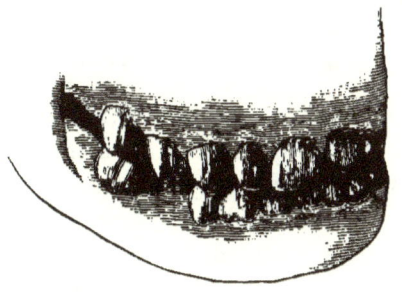

FIG. 146.

Case Corrected.

---
*Dental Cosmos, VOL. XXXIV., p. 19.

the deciduous molars below, a plate was made covering the upper posterior teeth, and so arranged that in addition to furnishing a masticating surface while the teeth were apart, it acted as an inclined plane in helping the lower jaw to move backward. From first to last he wore an occipito-mental sling, as illustrated in Garretson's Oral Surgery, increasing the tension from slight at first to as tight as could be borne without too great discomfort. At the end of nine weeks the articulation was normal, but the sling was worn for several weeks longer, without increased tension, to retain the satisfactory result secured. There can be no doubt that the wearing of a plate in the upper jaw arranged with an inclined plane, as described, will materially assist in forcing the lower jaw backward.

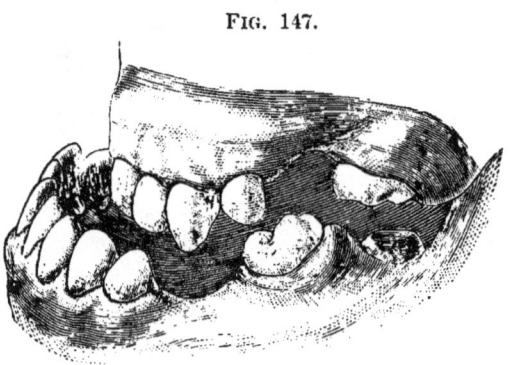

Fig. 147.

Excessive Prognathism.

Fig. 147 illustrates the most pronounced case of this class of deformities the writer has ever met with. The patient was a man of about forty years of age and was brought by a neighboring dentist for consultation as to whether anything could be done to remedy the defect. The lower jaw was very large in all its aspects, while the upper was correspondingly small. Although the lower incisors inclined decidedly inward, the distance from the cutting edge of the lower incisors to the cutting edge of the upper in a horizontal line, was a little over half an inch. From the upper jaw there were missing the right lateral, second bicuspid and first molar; while on the left side, the second bicuspid and two molars were absent. In the lower jaw, the patient had lost two molars and a bicuspid on the left side, and the first molar on the right. All the teeth of the upper jaw passed inside the lower, except the first bicuspids, whose external cusps articulated slightly with the anterior lingual cusps of the opposite molars below.

The advanced age of the patient, conjoined with the conditions just described, placed his case beyond surgical remedy and he was so informed. A plate covering and masking the natural teeth above with artifical teeth mounted outside to articulate with the lower ones was suggested, but the idea did not please him, and he concluded to pass the remaining portion of his life as he had the first, so far as his dental apparatus was concerned.

CHAPTER X.

LACK OF ANTERIOR OCCLUSION.

In certain rare instances, cases are met with in which the anterior teeth do not come in contact upon closure of the jaws. The bicuspids and molars of both jaws may articulate properly, but in the front part of the mouth there exists a space more or less great between the cutting edges of the incisors, when the jaws are closed. The space is greatest at the median line and gradually diminishes toward the cuspids. The condition not only gives a lisp to the speech of the individual, but renders these teeth entirely useless for purposes of mastication.

At first glance the incisors have the appearance of being too short in their crowns, but an examination will usually show that they are of normal size and length and that the process and possibly the maxilla itself is responsible for the shortened appearance.

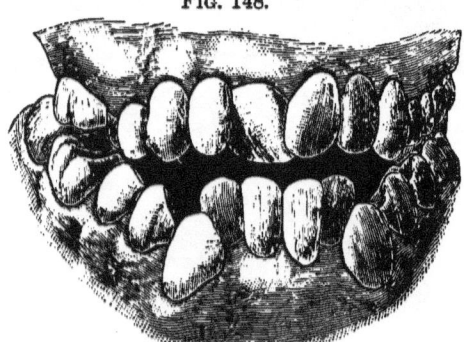

FIG. 148.

Lack of Anterior Occlusion.

In most cases it will be found that both arches are normal in form and size, that there is no antroversion or introversion either above or below, and that the superior teeth alone are at fault. Fig. 148 represents a typical case of this character, the model being from the collection of Dr. H. A. Baker.

Fortunately, the condition is seldom met with, for it is the one of all others that is least amenable to successful treatment.

The cause of the deformity has been variously attributed to thumb-sucking, to sleeping with the mouth open and to derangement of the articulation caused by ill-advised extraction of some of the posterior teeth; but while all of these are doubtless responsible for the condition in many instances, it is probably more frequently caused either by the lack of alveolar development in the incisor region, or an unaccountable variation in the plane of the alveolar border of the maxilla. The author has met with no cases of this condition that bore evidence of hereditary transmission, and therefore believes it to be due to a peculiarity in the development of the maxilla, originated with and confined to the individual himself.

One plan of treatment, where the deformity is slight, consists in grinding off the cusps and antagonizing points of some or all of the posterior teeth in order to shorten the bite and bring the anterior ones more nearly together. Much of this cannot be done without denuding the teeth of their enamel at certain points and exposing the sensitive dentine, but by grinding as much as is possible without causing too great pain and then administering an anæsthetic and continuing the grinding, quite an improvement can be brought about.

The sensitiveness of the exposed dentine may afterward be obtunded by repeated applications of either chloride of zinc, caustic potash or nitrate of silver. Where neither of these will avail sufficiently, it may be advisable to devitalize two or more of the teeth most interfering with occlusion and then continue the grinding until the necessary change is effected. The devitalized teeth will, of course, have to be subsequently treated and filled.

This method seems better than extraction because it leaves a portion of the crown for necessary mastication, but in some cases extraction may be the better plan.

Another plan, for aggravated cases, is to produce pressure upon the anterior portion of the lower jaw by means of a skeleton cap, chin-piece and rubber bands, very similar to the appliance used in retraction of the lower jaw, only that in the present case the power should be applied in an almost vertical direction. With such an apparatus, worn continuously for a few months, the condyles of the lower jaw will be tipped somewhat out of their cavities and the latter be partially filled up with new ossific material; at the same time the tendency will be to shorten the posterior occluding teeth by forcing them farther into their sockets and the correction in this way be incidentally assisted.

PART IV.

CHAPTER I.

CROWDED LOWER INCISORS.

While general consideration has been given to teeth erupting or situated inside or outside of the arch, there is one condition of rather common occurrence that calls for special mention. It is the crowded or jumbled condition of the inferior incisors after dentition is complete.

Fig. 149 shows an extreme case of this character.

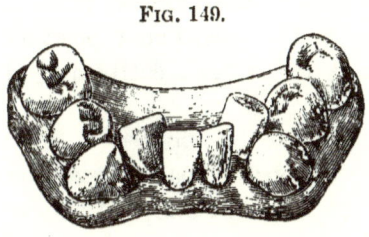

FIG. 149.

The moving of a single lower incisor either inward or outward into line has been treated of in Part III., Chapters I. and III., but where several or all of these teeth are more or less out of line and possibly turned upon their axes, the condition becomes a somewhat difficult one to treat successfully.

The expanding of the arch to permit of all of them being brought properly into line involves an operation of some magnitude, and is likely, in many cases, to disarrange an otherwise good articulation. For these reasons expansion should not be attempted except in very rare cases, where improvement of the articulation is desired and can be attained.

Two simple methods of treating these cases are open to us:—

First, where the crowding is not excessive, each of the

malposed teeth may be dressed off at the most prominent points of their approximal surfaces by means of hard rubber and corundum disks or by the more flexible emery-cloth disks, and by thus lessening their respective diameters proper accommodation may be found for them in the arch.

Second, where the teeth are very much crowded out of position and where the space between the cuspids is entirely inadequate for their accommodation it will be best to extract one of them in order to enable the remaining three to be brought into place.

As mentioned on page 51, the best tooth to extract in such cases is the one most out of line, or the one in such position as to enable the remaining ones to be most easily moved into proper alignment.

After extraction, means will have to be adopted to draw the teeth into position, in doing which the space created by extraction will also be closed. Perhaps the simplest way of drawing the teeth together is by the use of a rubber ring slipped over the teeth and kept from impinging upon the gum by a silk ligature wound several times around the terminal teeth near their necks and then tied to the ring itself.

After the teeth are drawn together and the space closed, they may be aligned by any of the appliances illustrated for moving individual incisors.

FIG. 150.

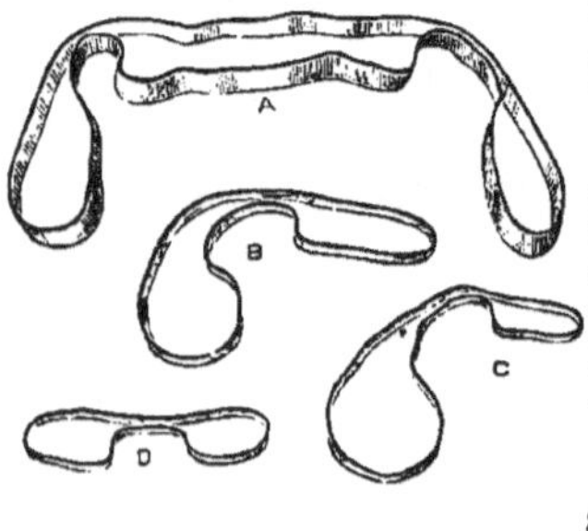

One of the best and simplest plans for moving outward the inferior incisors is by the use of the Byrnes metal strips, as shown in Fig. 150.

Appliance A was used to move all of the incisors, while B, C and D were employed to move individual ones.

Dr. Kingsley illustrates and describes an appliance and method for drawing the lower incisors together (after one has been extracted) and moving them into line at the same time; Fig. 151. He says:—* " It was a vulcanite plate with piano-wires, one from each side, meeting and lapping in front, and in their relaxed position standing off for an eighth of an inch from the face of the teeth, but were sprung in and tied to the incisors with waxed ligatures. This vulcanite plate was made pretty stout, comparatively non-elastic, and impinged upon the lingual walls of the bicuspids and molars, for the purpose of assisting nature, which was widening the arch by occlusion with the upper one, and, as from time to time it loosened by those teeth yielding, the plate was warmed and readjusted. A small ring from rubber tubing was also stretched over the three teeth, to assist in closing the gap. In four weeks the space was closed. The retaining fixture was exactly like the regulating plate without the piano-wire attachments."

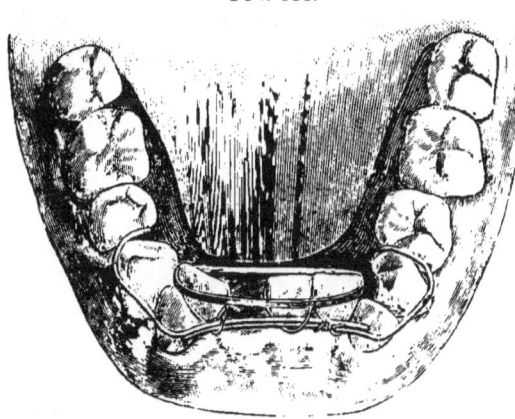

Fig. 151.

Another retaining appliance for cases of this character, occupying but little space and holding the teeth firm, is shown in Fig. 152. It is most conveniently constructed by fitting the bands to the

Fig. 152.

---

* *Dental Cosmos*, Vol. XXXIV., p. 106.

anchor teeth and then taking an impression of these in position, together with the lingual surfaces of the intervening teeth. A model of plaster and marble-dust, made from this impression, will enable us to fit the connecting strip and after securing it in place with binding wire, solder it to the bands.

Where greater accuracy is desired, the strip may be swaged up to shape by dies made from the same or duplicate model.

The retainer can be made to do a little delicate adjusting by slightly trimming on the model the still prominent corners of any teeth that we may desire to move outward.

CHAPTER II.

REDUCTION OF ELONGATION OF THE ANTERIOR TEETH.

Normally, each tooth will advance in the course of its eruption until the whole of its crown projects beyond the free margin of the gum, and its cutting edge or masticating surface is in proper relation with the same surfaces of the adjoining teeth. Full eruption may be delayed or entirely prevented, but extra elongation will not occur except through accidental circumstances. When it does occur, it is the result of an abnormal condition of the pericementum, most generally due to irritation in some form, or it is caused by lack of occlusion with teeth in the opposite jaw. In the latter case, it is but the manifestation of nature's attempt to rid the system of a useless organ.

Elongation of one or more of the superior incisor teeth sometimes occurs in connection with regulating and is due either to the irritation of the soft tissues surrounding the tooth caused by the impingement of the regulating appliance upon them, or to the unfortunate application of power in such manner as to favor the lifting of the tooth from its socket.

When such elongation is noticed it becomes necessary to remove the cause and give rest to the affected parts. The elongation being due in the first instance to the temporary thickening of the peridental membrane through irritation, a period of rest will usually result in the subsidence of the trouble and the return of the tooth to its former position. Where the elongation is the result of misdirection of power the operation will have to be suspended for a time, to be followed by the use of more suitable appliances. Should the condition, however, be allowed to continue for any length of time, as through non-appearance of the patient, some pressure may have to be applied to force the tooth back into its socket. This may be accomplished in a very simple manner by adopting the plan suggested by Dr. Wilhelm Herbst for retaining a replanted tooth.

It consists in cutting a short and narrow strip from a piece of rubber dam and perforating it in such manner that when in position, the crowns of two teeth on either side of the one affected will protrude through the openings, while the elongated tooth will be partly covered and pressed upon by the intervening portion of the rubber. Figs. 153 and 154 represent the strip of rubber separately and in position. Another way of producing tension upon the elongated tooth is by means of a rubber plate with a strip of gold so attached as to rest and press upon the cutting edge of the tooth.

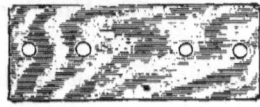

FIG. 153.

Herbst Method of Retention.

FIG. 154.

Rubber Strip Applied.

More elaborate in character, but well adapted to the purpose, is the appliance devised by Dr. Dodge.*

The right central was elongated and needed reduction. A double-cap was made of gold to fit and cover the left central and lateral and a similar one to cover the right lateral. These caps were joined on their labial and palatal surfaces by heavy gold wires, each having three headed pins attached to it. When completed, it was cemented in position, as shown in Fig. 155.

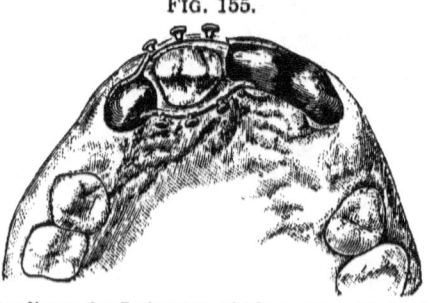

FIG. 155.

Appliance for Reduction of Elongation. (Dodge.)

The appliance for furnishing

---

* *Dental Cosmos*, Vol. XXXIII., p. 1045 et seq.

the power consisted of a short piece of elastic braid, to one end of which were attached three small gold rings and to the other a gold bar fitted with tube, ring-bolt and nut, as seen in Fig. 156.

In use the three gold rings engaged with the three pins on palatal bar while the ring-bolt was caught over the central pin on labial bar. The elasticity of the braid, after being lessened by the moving of the tooth was increased by turning the nut on the ring-bolt.

Fig. 156.

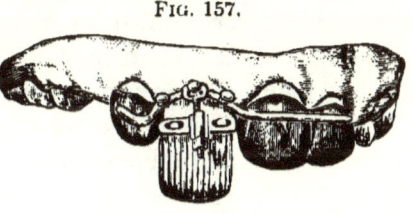

Fig. 157.

Appliance in Position. (Dodge.)

The completed fixture in position is shown in Fig. 157.

None of the appliances mentioned need be worn long, for the continuous pressure will quickly cause recession of the tooth.

CHAPTER III.

ASSISTED ERUPTION OF THE ANTERIOR TEETH.

Incisor teeth that have not erupted to their full extent and have been prevented from doing so by too close proximity of adjoining teeth or other cause, may often be assisted in assuming their proper alignment. Where space exists, teeth will naturally accomplish their full eruption unaided, as previously stated. When they do not, and there is no visible cause for their not doing so, we may safely infer that some hindrance exists in the tissues beneath the gum. It may only be an unexplainable suspension of the act of eruption, or it may be, and often is, a curvature or enlargement of the root that prevents the further progress of the tooth. Which of the two it is, can usually only be decided after measures of assistance have been tried.

If the delayed eruption has been due simply to a suspension of the act of eruption, the simplest and most effective remedy will be found in tying a silk ligature around the neck of the tooth and pressing it well under the free margin of the gum, or in placing a ring cut from rubber tubing in the same position. Either one will cause irritation of the pericementum, which by consequent enlargement will tend to force the tooth out of its socket. To prevent undue elongation the case will have to be carefully watched, day by day, and the irritating ligature removed as soon as the tooth has been sufficiently elongated. Should this be neglected, the tooth might be entirely expelled and lost.

Should these simple means fail to move the tooth from its abnormal position, osseous abnormality is probably the hindering cause, and mechanical appliances of not too great power should be tried. Some of this character have been mentioned in Part III., Chapter 2.

Dr. A. E. Matteson* has devised an appliance for producing forced elongation of several of the incisor teeth at the same time. It is composed of a rubber plate to which a piece of clock spring, properly shaped, is attached. The spring is cut and ground along its outer edge in such manner as to leave projections to pass between the teeth at their necks and bear upon the wider parts of the crowns. After being properly shaped and fitted, the spring is riveted to the anterior portion of the plate just back of the teeth to be acted upon. In inserting the appliance, the projections of the spring are passed between the teeth at their necks and the plate pressed into place. The elasticity of the slightly curved spring with its projections, will produce pressure upon the teeth in the direction of their length and cause their elongation.

The action of all appliances of this character will have to be closely watched to see that the force exerted by them is neither too great nor too long continued.

Should any or all of the appliances mentioned fail to move the partially erupted tooth, we may safely conclude that its root is exostosed or curved at some point of its length, and further operations had better be suspended.

The author, in his early practice, attempted to rotate a superior cuspid tooth, and after failing to produce any effect by the commonly adopted appliances, concluded that the trouble must lie in the formation of the root. A digital examination of the tissues overlying the root, revealed the fact that it was considerably curved, and further efforts at rotation were immediately abandoned. Had the examination been made before beginning operations, as it should have been, instead of at their close, much annoyance and trouble would have been spared both patient and operator.

Where full eruption of a tooth has been made impossible by the impingement of adjoining teeth upon the space intended for it, increase of space by lateral pressure upon the

---

* Harris' Principles and Practice, 12 Ed., p. 439.

interfering teeth should first be gained before any attempt is made at elongation. Indeed, the mere enlargement of the space and its retention for a length of time will usually be followed by the unaided eruption of the tooth. Should this not occur, mechanical assistance may be rendered by some of the methods mentioned.

Forcible eruption of a tooth by means of the extracting forceps is seldom justifiable, for we cannot always know what may have interfered with the eruption. In certain exceptional cases, where a careful examination reveals no sign of malformation of the root, and where it is perfectly evident that slight impingement of adjoining teeth has been the sole hindrance to full eruption, the forceps may prove a valuable means of effecting a rapid and easy correction of the difficulty.

Such a case occurred in the author's practice. The patient was a gentleman of about twenty-eight years of age, whose right central incisor was about a line shorter than its mate. It had been tardy in erupting and in consequence there was a slight lack of space for its accommodation, as shown in Fig. 158.

Fig. 158.

Incomplete Eruption.

As the difference in length of the two incisors was too great to be remedied by the simple means of reducing the length of the longer one, it was decided to elongate the shorter one. A careful examination proving favorable, a piece of sand paper was folded so as to cover both labial and palatal surfaces of the tooth to protect it from injury, after which it was grasped with the forceps and by a combined rotary and downward motion brought into place. Once in position, it was held there firmly by the pressure of the adjoining teeth, but as good judgment would not sanction so unreliable a means of retention, an appliance had to be devised that would not only prevent the tooth from slipping back into its socket but also secure it from being forced forward by pressure upon its sides. The patient also desired the appliance to be as inconspicuous as possible.

To accomplish all of these ends, a piece of platinized gold wire, a little thicker than a vulcanite tooth-pin, was bent into horseshoe form and curved to conform to the palatal surfaces of the retarded tooth and the two adjoining ones. The ends of the wire were then flattened and bent so that they would hook over and rest upon the cutting edges of the adjoining central and lateral. A silk ligature was passed around the moved tooth and tied in front, after which the ends were again passed to the palatal surface and tied just below the cingulum. After the gold wire was placed in position, the ligature was attached to it at the lowest point of its central curve.

The ligature thus held the appliance in position and it in turn kept the tooth from receding. The double arrangement of wire and ligature also guarded the tooth against the possibility of moving forward. The fixture in position is shown in Fig. 159. The only parts of it visible were the small rounded gold tips that overlapped the cutting edges of the two adjacent teeth.

Fig. 159.

Retention after Correction.

Where sufficient space exists for the purpose, the tooth after being drawn into position may be held by means of the platinum band and extension bar, as shown elsewhere for retaining a tooth that has been forced backward into the line of the arch.

CHAPTER IV.

TOOTH-SHAPING.

During the act of regulating or after its accomplishment, one of the most useful accessory operations, when called for, is that of dressing or shaping certain teeth so as still further to improve their appearance.

This operation will probably not be necessary in the majority of cases we treat, but when indicated, it adds very much to the patient's appearance and the satisfaction of the parent and operator. It may be accomplished by means of the file, corundum point, sand-paper disk or emery cloth strips, each having value according to the requirements of the case.

It will not often be called for on the approximal surfaces of teeth, but when it is, much of the substance should not be removed and the surface should afterward be polished in the most perfect manner.

The author has had one case, and one only, in which such trimming of approximal surfaces seemed advisable. The patient was a young lady of about twenty-one years of age, whose anterior superior teeth were slightly prominent. The teeth were without interdental spaces and all of the posterior ones were so perfect in structure, alignment and occlusion, that the extraction of even one of them would have been regarded as an unwarrantable sacrifice.

All of the six anterior teeth had small cavities upon each of their approximal surfaces, and it was therefore decided that in the filling of these cavities a slight portion of each approximal surface should be dressed off in the hope that the aggregate of such spacing would be sufficient to enable the teeth to occupy a position more in harmony with the normal line of the arch. After the filling and dressing of the surfaces, the teeth were drawn inward and the result was all that could have been desired.

Sometimes teeth that have fully erupted out of line, when brought into proper position extend below the line of the cutting edges of their neighbors and the rest of the teeth in the arch. Any attempt to reduce their elongation by forcing them up into the socket would not only be extremely difficult, but in many cases futile. The better plan, if the disparity in length be not great, is to grind off their cutting edges somewhat, and thus accomplish the desired end in a very simple manner.

Fig. 160 shows a case of this character, and Fig. 161 the improvement after grinding.

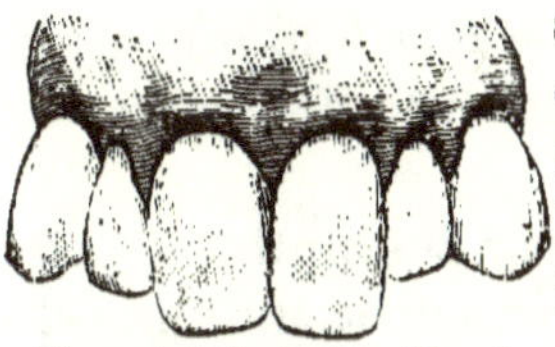

FIG. 160.
Elongated Centrals. (How.)

FIG. 161.
Improvement by Grinding.

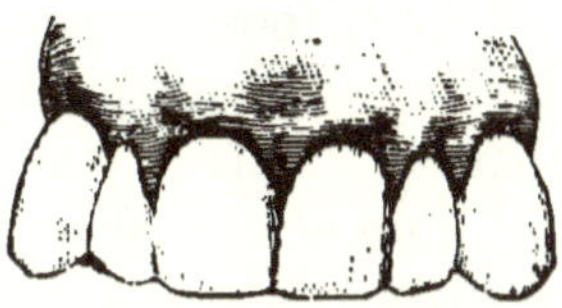

Again, teeth out of line have from lack of attrition preserved their normal, rounded form, while their fellows have been more or less worn away on their cutting edges either through abnormal occlusion or excessive use. When the malposed teeth have been brought into position their rounded and unworn cutting edges are apt to contrast strongly with the abraded edges of their neighbors. By so dressing the incising edges of the unworn teeth as to resemble those next to them, greater harmony of expression will result.

FIG. 162.
Unevenly Worn Incisors. (How.)

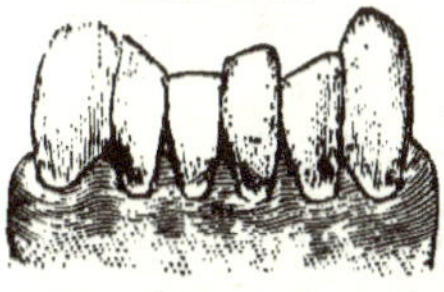

FIG. 163.
Improvement.

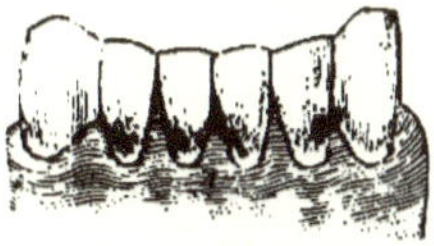

Fig. 162 illustrates a case in which the worn condition and varying length of the lower anterior teeth presented a very unsightly appearance, and Fig. 163 represents the improvement made by reducing the length of certain ones and straightening the edges of others by grinding.

Altering the form of a tooth, however, may often be made to serve even a more useful purpose than that of appearance. Cases have occurred where an upper tooth, tardy of eruption, has been unable to come entirely down into line owing to the meeting of its antagonist of the opposite jaw edge to edge. In such an event, the retarded tooth might be forced sufficiently outward to enable it to accomplish its full eruption and then be held in position until overlapping had taken place, but the operation may be advantageously simplified in most cases by slightly beveling the edge or cusp of the lower tooth on its labial, and the upper one on its palatal surface. The inclined plane thus formed will enable the upper tooth to slide over the lower one into line, which it will be almost certain to do provided there be no contingent obstructions.

A case of this character came under the author's notice recently in which a superior lateral incisor was thus impeded in eruption until the individual was forty years of age. A simple beveling of the cutting edges of it and its opponent, induced it to come into proper line within a year.

Other conditions than those just mentioned will occur to the practitioner in which the slight alteration of the form of a tooth will materially assist, or be the means of entirely accomplishing some simple act of regulating, and in other cases, greatly add to the effect of some long-continued and otherwise successful operation in orthodontia.

CHAPTER V.

CONSTRUCTION OF REGULATING APPLIANCES.

The principal tools required for the construction of metal regulating appliances are illustrated in Plates I. and II.

"a" is the ordinary mouth blow-pipe to be used in connection with a large alcohol annealing-lamp or gas Bunsen burner; "b" is the best form of jeweler's pin-vice, having pivoted jaws operated by an inclined plane on revolving handle. The handle is bored entirely through to receive wire of any length. "c" is a "snip" plate shears of the form recently introduced for crown- and bridge-work. "d" is a dental pin-punch and "f" a solder tweezers. "e" is known as a clasp-bender, and with its one beak of cylindrical form and the other concave is a powerful and useful instrument for curving and shaping piano or other stiff wire.

"g" and "h" are respectively flat and round nose pliers, while "i" is a heavy pliers for drawing wire or tubing, with notches in the joint for cutting wire. "j" is a small steel anvil mounted in a metal base, and "m" is a jeweler's saw-frame and saw. "k" is a small jeweler's set of taps and dies for cutting threads in nuts or upon wire, while "l" is a screw plate, (Stubs No. 19) usually accompanied by corresponding taps for the same purpose. "n" is a cut-nippers, and "o" a pair of contouring pliers, as used in crown-work. "p" is a copper soldering iron for soft soldering, and "q" and "r" metal gauges. "q" is the Standard American Gauge, very generally used for both plate and wire.* "r"

---

*Inasmuch as confusion has arisen at times in describing the different thicknesses of wire and plate used in dental work, some writers using the English Standard Wire Gauge (which is similar to the Birmingham and Stub's gauges), others the American Standard Wire Gauge (sometimes called the Brown and Sharpe or B. & S.), while nearly all number piano wire acis a micrometer gauge for more delicate and accurate measurements, which are indicated in thousandths of an inch. "*s*" is a double-calliper, with one end for inside and the other for outside measurement.

"*t*" is a self-closing tweezers for holding parts in soldering, and "*u*" a draw-plate, known as the "Joubert," with thirty different sizes of holes. The illustrations "*v*" to "*z*" will be referred to in the description of processes. "*aa*" represents a metal ferrule or band, the edges being drawn tight and held together with fine iron binding-wire. "*bb*," "*cc*" and "*dd*" are simple forms of wire clips for holding parts in position while soldering. They are made from piano-wire, Nos. 17 to 21, each about an inch in length. In "*bb*" both ends are beveled inward to enable them more easily to slip over the parts to be held. In "*cc*" one end of the clip is formed into a loop and the other bent at a right angle with flat termination. "*dd*" is a modification of "*bb*," one arm being curved near its end to hold a tube in parallel position upon a band.

"*gg*" is intended to represent the manner of holding a tube at right angles to the length of the band. The wire clip for this purpose has one arm bent at a right angle near its end and flattened and made concave to fit the tubing, while the other arm is straight, as in "*bb*."

"*hh*" represents a simple wire support for holding parts in position upon a soldering block, as in constructing a gold T. The support is made from a piece of piano-wire flattened and drilled at one end, tapering to a point at the other and bent at a right angle near the middle. The pointed end is easily forced into an asbestos or charcoal block at any desired place.

---

cording to the Steel Music Wire Gauge (Washburn & Moen Co.), the author would suggest the adoption by American dentists of the American Standard Gauge (B. & S.) for indicating the thickness of all plate and wire used for dental purposes. The numbers indicated in this chapter refer to the B. & S. gauge. A comparative table of the various gauge measurements, by which the equivalent of one may be found in any other is appended to this chapter.

PLATE I.

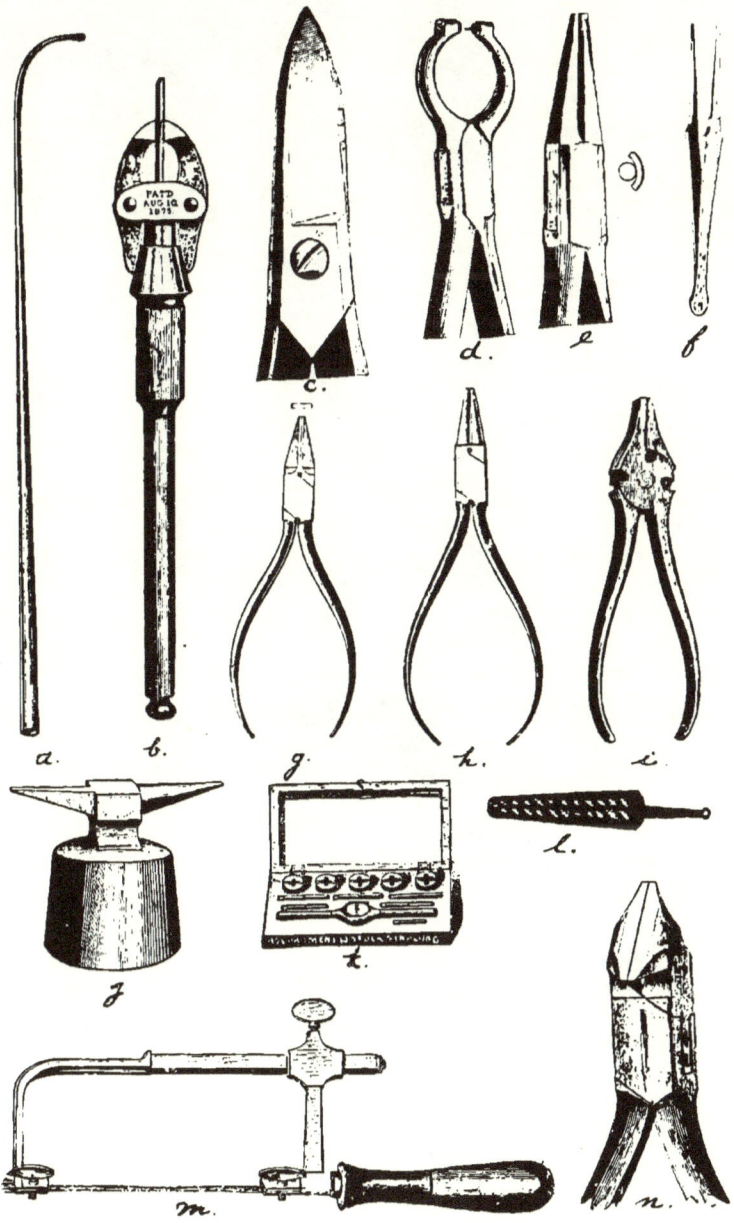

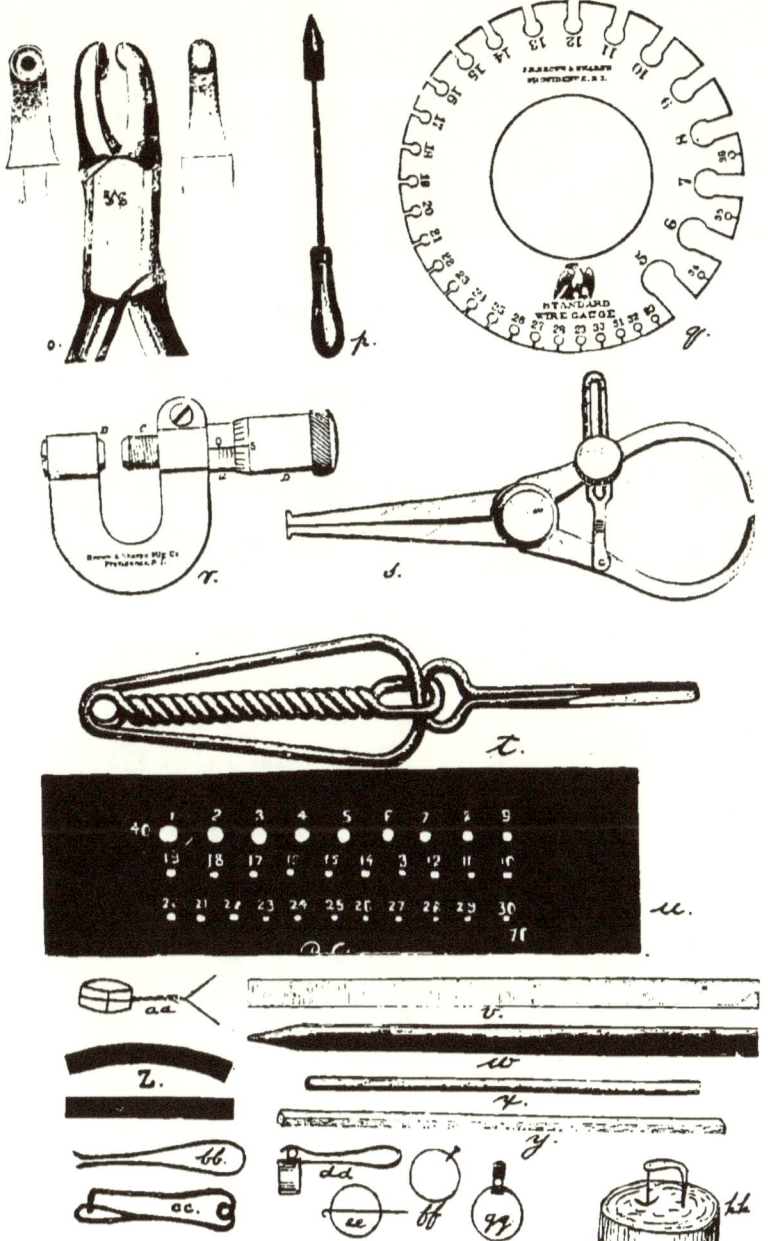

PLATE II.

FERRULES OR BANDS.

Ferrules or bands for encircling teeth and serving as means of attachment for operating or retaining appliances may be made from gold plate (18 to 22 karats fine), platinized gold, iridio-platinum, platinum, platinized silver or German silver. All of those mentioned, except the last, will remain nearly free from oxidation, but German silver soon becomes dark in the mouth. As a rule, bands should be made to fit loosely so as to afford slight space for the cement which is to hold them in position, and where practicable, the tooth to be fitted should be freed from contact with its neighbors by previous wedging. As this cannot always be done, the bands in some cases will have to be forced over the teeth in spite of their contact. In such event they should be constructed from the stiffest and least yielding of the metals mentioned, such as platinized gold, platinized silver, or iridio-platinum. Bands made from these metals, even though thin, will retain their form without "buckling" while being forced into place. Where the teeth to be banded are not in close contact, any of the other metals will serve as well for the construction of bands. The band material should not exceed No. 32 in thickness and should be cut into strips from $\frac{4}{32}$ to $\frac{5}{32}$ of an inch in width.

For the six anterior teeth the strips should be curved as shown in "z," so that when bent to encircle the tooth with the convex edge toward the gum, the ends will meet one another or overlap on the lingual surface in a nearly horizontal line.

For molars and bicuspids the band should be straight, and if desired, may be contoured transversely with the contouring pliers (o).

Ferrules are neatest when made with a flush joint, and their ends may be held in close apposition by passing binding wire around them (aa) and twisting the ends to serve as a holder while soldering in the flame of a lamp. When made with a lap joint they may be secured in the same way or the lapped ends may be held with a wire clip (*bb*).

In some cases, as in partly erupted cuspids and deciduous molars, where the exposed portion of the crown is short and conical, it is important to have the band more accurately adapted to the form of the tooth so as to gain a firmer hold.

This may easily be accomplished by reproducing the crown in Melotte's metal, and after roughly adapting the band to the natural tooth or its duplicate in plaster, completing the operation on the metal crown with hammer and chaser, or by swaging it between a die and counter. Prof. Matteson prefers the latter plan, the results of which are nicely shown in Fig. 65, p. 123.

The attachments that bands are most commonly supplied with are headed pins, wire hooks and pieces of metal tubing. Tubes are held in position upon bands for soldering by means of wire clips as shown in "*dd*" and "*gg*."

As tubes are usually not soldered along their edges after being drawn, they can be closed at the time of soldering to the bands by placing the joint next to the band, and when desired to be left open the joint is turned away so as not to be included in the soldering.

Headed pins may be obtained from a vulcanite tooth and after being filed to a point can be inserted into a hole drilled in the band, as shown in "*ff*."

When a band is to be supplied with hooks on opposite sides, a convenient way of attaching them is to drill a hole in each side of the band and pass entirely through them a wire bent into a hook at one end as shown in "*ee*." After soldering, the straight end is also bent and the wire cut from the centre of the band. A hook for one side only may be inserted and held like the pin in "*ff*."

ROUND TUBING.

Tubing for pipes or tubes can usually be obtained at jeweler's supply houses. It may be had of gold, brass or German silver or of one of the latter metals plated. It comes in lengths of about four inches, and is smoothly drawn but not soldered (*x*).

The thickness of the tubes is generally greater than we desire, but after soldering the joints with silver solder, the inside diameter can be enlarged with an engine bur, a spear or fissure drill, or jeweler's reamer. In many cases it may be desirable or necessary to manufacture our own tubing, which can be done as follows: Select a piece of metal plate of suitable gauge (No. 27) and cut from it a strip of desired length and of a width equal to three and a third times the outside diameter of the proposed tubing (*r*). Shape one end of the strip like the nib of a pen and curve or round the entire piece somewhat by forcing it into a groove cut in a block of hard wood, using a piece of wire and hammer for the purpose (*w*). The pointed end is then passed into one of the larger holes of the draw-plate, seized with the pliers (*i*) and drawn through. This operation is repeated through the holes next in size until the cut edges of the strip are in close apposition. If it be desired to reduce the external diameter after the tube is formed, it can be done by simply continuing the process. As the drawing stiffens the metal it will be necessary to anneal it occasionally during the process. Where the tubing is to be used in considerable lengths without soldering, where great stiffness is required, as in the Angle encasement for jack-screws, there should be no annealing of the metal near the close of the operation of drawing.

SQUARE TUBING.

This form of tubing can seldom be bought, and so will have to be manufactured. The strip will first have to be converted into a round tube, as previously described, and drawn to near the proper size in the ordinary draw-plate, after which the last four or five drawings must be made through a square hole draw-plate. This form of tubing (*y*) when made of plate a little heavier than usual, and drawn to a size to fit the wrench or key, is admirably adapted for the construction of small nuts as well as heads of screws.

Sections of any length can be cut from it with a jeweler's saw, after which by grasping them in a hand-vise (b) they may be readily drilled or reamed to proper size and tapped. The saw should always be set in the frame with the teeth pointing toward the handle so that the cutting is done in drawing the saw backward. Reversing the operation would cause the back of the frame to spring and the saw be liable to break in consequence.

WIRE-DRAWING.

The process of drawing wire down to size is very similar to that of drawing round tubing. Steel wire cannot be drawn by hand, but can be bought in all sizes. Gold, platinum or German silver wire, however, can readily be reduced in diameter and correspondingly lengthened by means of the draw-plate (u) and the heavy pliers (i). Before using the draw-plate, the holes should be filled with melted bees-wax or equal parts of bees-wax and tallow to act as a lubricant. It is then clamped firmly in the bench-vise. The wire after being annealed should be reduced at one end by filing or hammering and the pointed end passed through the hole in the draw-plate next less in diameter than the wire itself. It is then grasped by the pliers (i) and drawn through with a continuous and steady pull. In similar manner it is drawn through the successively smaller holes until the desired gauge is obtained. After each three or four drawings the wire should be annealed, otherwise its increasing brittleness may cause it to break.

Bending wire.—Wire may be bent into any form by means of the various pliers, assisted at times by the bench- or hand-vise. Any curve can be given to it with the round-nose pliers (h), while for bending it to a right or acute angle, it should be held in the vise or pliers close to the desired place and the free end grasped and bent over by the flat-nose pliers. If the wire be of large size it may best be bent to a right angle by grasping it in the bench-vise and forcing the free end down with a hammer.

When piano wire needs a sharp bend it should always be done in this latter manner. For bending piano wire into a short curve, as for making or altering the form of the Coffin W spring, the clasp-benders (e) should be used, on account of their convenience and superior power.

SOFT-SOLDERING.

In uniting small parts of appliances by means of soft-solder, they may be held in the tweezers (f), (t) or the spring clips bb, cc, dd, or wrapped with binding wire. After applying the soldering fluid, the piece is held over an annealing lamp and when sufficiently heated is touched with the end of a thin rod of solder which at once melts and unites the parts. In this way the minimum amount of solder may be applied.

For soldering larger parts of appliances, as in forming the Jackson cribs and springs, they should be secured in proper position on the plaster model, and after applying the fluid and laying a piece of solder on the parts, the latter is melted and the parts joined with the soldering iron (p), previously heated over a Bunsen burner. In soldering steel (as piano-wire) the fluid causes oxidation of the metal so quickly that it is important to heat the parts and melt the solder immediately after the fluid is applied.

HARD-SOLDERING.

Both the student and practitioner are supposed to be familiar with this process, so that but few suggestions will be needed.

Parts of German silver appliances should be soldered with silver solder (silver 2, brass 1) while any of the compounds of gold or platinum may be united with either silver or gold solder. The latter when used should not be less than 18 karats fine to keep its color in the mouth. In joining articles with hard solder the parts to be united should be touched with the least quantity of liquified borax and only as much solder applied as is necessary.

After drying with moderate heat, the full flame should be directed upon the parts to be united and fusion accomplished as quickly as possible. Most of the hard-soldering required in constructing portions of regulating appliances may be done by holding the parts in the flame of a small Bunsen burner or alcohol annealing lamp. In other cases a larger flame with a blow-pipe to direct and concentrate it will be necessary, the parts being laid or secured in position upon a soldering block made of charcoal, asbestos, pumice-stone or other suitable substance.

When two solderings are necessary for the same piece, the first joint may be kept from unsoldering during the second process by placing an extra wire clip upon it or by seizing it with the tweezers at that point and thus protecting it from over-heating.

In soldering bands together as in Fig. 19, they may conveniently be held by means of the clip " *bb*," while a band and bar, as in Fig. 17, may be held in the same manner.

When two pieces of tubing are to be united at other than a right angle, they may be arranged and held in position upon the soldering block by being pinned down with staples made from piano wire. The end of one piece will of course have to be filed concave to fit the convex surface of the other, before joining.

Where it is desired to lengthen a traction screw, it may be done with less labor than constructing a new one, by cutting it in two at some point, inserting a new piece of like diameter, and uniting with solder. The parts may be held in position by being pinned to some smooth surface of the soldering block, or a groove may be made in the block with a straight piece of wire and hammer and the parts to be united laid into it. A wire may be shortened in like manner.

When a wire or a tube is to be joined to another at a right angle, they may be held as shown in "*hh*," which is also one of the best means for holding two flat pieces of metal for soldering, as in constructing a T appliance.

A screw cut wire may be fitted with a square end for turning with a key, by filing it to a smaller diameter and slipping over it and soldering a section of square tubing. If a shoulder is wanted in addition, a small washer or disk may be placed on the wire before the square tubing is adjusted. Both will be united at the same soldering.

SCREWS AND NUTS.

The making of jack- or traction-screws and nuts to play upon them is an operation requiring care, but is not beset with much difficulty.

A set of taps and dies (k) or a screw-plate (l) with corresponding taps, can be procured at jeweler's supply houses or from dealers in fine tools, and they, with a bench or pin-vise and a few tools such as files, and pliers, together with a few sizes of jeweler's reamers will furnish us with all the equipment we need for the purpose.

Screws.—The wire that is to be threaded or screw-cut, should be smoothly drawn and of moderate temper. The form that one end of the wire is to have may be given to it either before or after the thread is cut upon the other, but in most cases it will be well to fashion it first. The wire should be of exactly the same diameter as the hole in the screw-plate would be if the threads were removed. If smaller than this, a full, deep thread will not be cut, and if larger, the wire will be liable to be twisted off in the attempt. In cutting the thread, the wire should be slightly tapered at its end and grasped in the bench-vise, horizontally or vertically, with only about half an inch of its length protruding. To avoid marring the wire, the jaws of the vise should be provided with lead or brass caps.

The screw-plate (or die) should now be held against the end of the wire at right angles to it and given a quarter or half turn with firm pressure. This should be repeated four or five times until the tool is well started in its work, care being taken to preserve the forward pressure and to see that the screw-plate is kept at a right angle to the wire. The operation may now proceed more rapidly until all of the exposed portion of the wire has been covered.

If a longer portion is to be threaded, more of the wire may now be exposed and the operation continued. A little oil fed to the tool will greatly facilitate the cutting. In reversing the operation to release the tool, care should be exercised not to mar the thread.*

Nuts.—For use in regulating appliances, nuts should be square and about No. 13 in diameter so as to fit an ordinary watch key or wrench and should not be less than $\frac{3}{32}$ of an inch in length in order to have a good hold and resist the necessary strain. When greater strain than ordinary is to be withstood, they should be made longer to prevent stripping of the thread. They may be made from a nickel five-cent piece by filing it smooth on both sides, marking into squares of proper size, centering, drilling and tapping each one and then sawing them apart, as recommended by Dr. Case; or they may more conveniently be made from heavy, square, German silver tubing, by sawing it into sections, grasping these in the pin-vise (b) and then drilling and tapping them.

In using the tap to cut the threads on the inside of these nuts, it should be held in a suitable holder and fed carefully with an alternate forward and backward movement to avoid clogging and danger of breaking the highly tempered tool.

The taps and dies are always marked with the same numbers so that the corresponding ones can be used in each case. If the tubing from which the nuts are to be made has not been soldered after being drawn, it should be before it is cut into sections.

---

* German silver wire of any size can be procured at electrical supply houses. The size best adapted for jack- and traction-screws is No. 17, (B. & S.) corresponding to hole No. 17 in the *Joubert* draw-plate. This wire is of just the proper diameter to be thread-cut in the third smallest hole of Stub's screw-plate (*l*).

PRACTICAL TREATMENT. 221

Comparative Table,

SHOWING SIZES OF WIRE AND PLATE IN DECIMALS OF AN INCH BY VARIOUS WIRE GAUGES.

No. of Gauge.	English Standard.	Stub's and Birmingham.	American Standard or Brown & Sharpe.
0000	.454	.454	.460
000	.425	.425	.40964
00	.380	.380	.36480
0	.340	.340	.32495
1	.300	.300	.28930
2	.284	.284	.25763
3	.259	.259	.22942
4	.238	.238	.20431
5	.220	.220	.18194
6	.203	.203	.16202
7	.180	.180	.14428
8	.165	.165	.12849
9	.148	.148	.11443
10	.134	.134	.10189
11	.120	.120	.09074
12	.109	.109	.08081
13	.095	.095	.07196
14	.083	.083	.06408
15	.072	.072	.05706
16	.065	.065	.05082
17	.058	.058	.04525
18	.049	.049	.04030
19	.040	.042	.03589
20	.035	.035	.03196
21	.0315	.032	.02846
22	.0295	.028	.02534
23	.027	.025	.02257
24	.025	.022	.0201
25	.023	.020	.0179
26	.0205	.018	.01594
27	.01875	.016	.01419
28	.0165	.014	.01264
29	.0155	.013	.01125
30	.01375	.012	.01002
31	.01225	.010	.00892
32	.01125	.009	.00795
33	.01025	.008	.00708
34	.0095	.007	.0063
35	.009	.005	.00561
36	.0075	.004	.005

In explanation of the above table it may be stated that the numbers in the first column refer to the numbers on the various gauge-plates, and their respective equivalents, in decimals of an inch, will be found in one of the other columns. Thus No. 17 of the B. & S. gauge is equal to .04525 of an inch, while No. 17 of the English standard is equal to .058, or a trifle more than .010 greater than that of the B. & S.

So also No. 18 of either the Stub's or English gauge is nearly the same as No. 17 of the B. & S.

CHAPTER VI.

ELECTRO-PLATING.

Electro-plating is the art of precipitating certain metals from their solutions by the slow action of a galvanic current. By this process the salts of the metals in solution are decomposed, the metal being deposited upon the object to be plated at the negative pole while acid is liberated at the positive one. Electro-gilding, or plating with gold, is employed in dentistry chiefly for the purpose of giving to appliances made from the baser or oxidizable metals a coating of finer metal (gold) that will resist the action of the fluids of the mouth.

Regulating appliances made from German silver or steel, in whole or in part, not only present a better appearance, but endure longer and operate more satisfactorily when properly gilded.

Piano-wire, so valuable and so largely employed in connection with regulating devices, not only becomes unsightly, but deteriorates when worn for a long time in the mouth. It may be bought electro-gilded, but the coating is so thin as not to be durable.

German silver, which is rapidly growing in favor for the construction of regulating appliances on account of its inexpensiveness and intrinsic merits, is also readily acted upon by the fluids of the mouth and the resultant oxidation greatly interferes with the operation of nuts and screws or the play of wires in their neatly fitting tubes. Gilding obviates all of these disadvantages and gives us the virtue of gold without its expensiveness.

German silver is easily gilded in either a warm or cold bath if its surface be first thoroughly cleansed, but steel, owing to its ready oxidability in the cleansing bath, does not receive a good and durable coating of gold unless it is first plated with copper. Steel therefore requires to be subjected to two processes, while German silver or other alloys of copper need but one. Each of these processes will be described.

As a preliminary to plating, all articles must have a perfectly clean surface, otherwise the deposit will not adhere firmly to the object receiving it and the durability of the coating be greatly lessened.

During both processes of cleansing and plating, the article must not be touched with the fingers, as the slightest contact will prevent the adhesion of the metal at such points. To prevent this the article to be plated should have a copper or platinum wire attached to it at some point by means of which it must always be handled until the entire operation is completed.

CLEANSING GERMAN SILVER, BRASS AND OTHER COPPER ALLOYS.

The following is one of the best formulae for a cleansing solution:

Caustic potash,	1 lb.
Water, (soft)	1 gal.

Heat nearly to boiling point in a glass, porcelain or porcelain-lined dish, and suspend the article for a few minutes in the hot solution.

Remove and brush thoroughly upon a board, after which rinse well in clean water. If the article is soft-soldered at any point it must not remain in the lye too long or the solder will be acted upon.

CLEANSING STEEL.

Dip in the caustic lye used for copper, rinse thoroughly, scour with moistened pumice, rinse again and pass through the following dip:

Sulphuric acid,	1 part.
Water,	20 parts.

After this the article must again be well rinsed before being placed in the plating bath.

COPPER SOLUTION.

The electro deposit of copper is usually obtained by the decomposition of acetate of copper and cyanide of potassium.
A good bath or solution is as follows:

Water, (soft)	1 gal.
Acetate of copper, (crystals)	3½ ozs.
Carbonate of soda, (crystals)	3½ ozs.
Bisulphite of soda,	3 ozs.
Cyanide of potassium, (pure)	7½ ozs.

Moisten the copper salt with water to form a paste, (otherwise it is apt to float on the liquid); stir in next the carbonate of soda with a little more water, then the bisulphite, and finally the cyanide with the rest of the water.

When solution is complete, the liquid should be nearly colorless. If not, add cyanide until it is.

This bath may be used either hot or cold.

An immersion of a few minutes will usually furnish a sufficient coating of copper when the article is afterward to be gilded.

GOLD SOLUTION.

Formula:

Chloride of gold,	72 grs.
Cyanide of potassium, (pure)	1½ ozs.
Water, (distilled)	30 ozs.

Dissolve the cyanide in part of the water, then gradually add the gold chloride* dissolved in the remainder. Boil for one-half hour and use cold. The solution prepared as above should be colorless after standing awhile, and the color of the deposit should be yellow. If black or dark red, add more cyanide dissolved in water. If cyanide be in excess, plating will proceed slowly or not at all; in such case add more gold chloride or increase intensity of current by im-

---

* Chloride of gold can be purchased at chemical supply houses.

mersing zincs deeper in cell. In gilding German silver, the best results are obtained when the bath is kept slightly warm.

All gilding baths should be stirred occasionally to destroy the gravity of the liquids.

BATTERY.

For small articles, such as regulating appliances, a single cell (Daniel or Smee) will give us sufficient intensity of current.

Fig. 164 represents a simple battery composed of a single Smee cell connected with the jar containing solution and articles to be plated.*

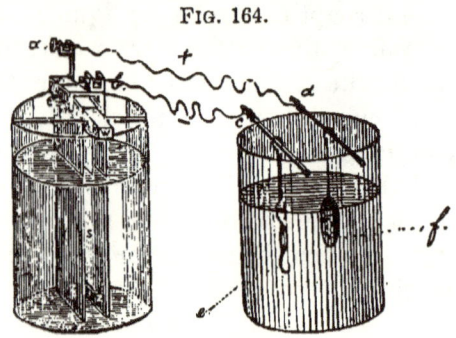

Fig. 164.

Plating Battery.

In the illustration of the battery, "z" represents the zinc plates and "s" a strip of platinized silver suspended between them. "w" is a wooden support which rests upon the edges of the jar with the silver strip let into its under surface. "b" is a clamp connecting the two zinc plates on the outside of the wooden support, while clamp "a" connects with the silver strip. The plating or bath jar has two copper rods resting upon it, one of which "d" has the metal anode suspended from it by means of a platinum or copper wire, while the other "c" has the cathode or article to be plated similarly suspended.

The battery jar or cell is charged with a solution of one part of sulphuric acid to ten parts of water. The current

---

* A Smee cell with platinized silver plate can be bought at any electrical supply house for $3.50, and additional plain glass jars, one for copper and the other for gold solution, for about 15 cents each.

generated by the action of the acid solution upon the zinc plates passes through the positive (+) wire to the rod "d," into the plating solution by way of the anode (f), across to the cathode (e) and back to the battery by way of the negative wire (−). The current in its passage decomposes a portion of the gilding solution and causes the metal thus set free to be attracted to and deposited upon the article to be plated.

The solution being thus deprived of a portion of its constituent salt, in turns acts upon the suspended anode and takes from it a sufficiency of the metal to restore its former equilibrium. In this way, as each article is coated with the metal, the suspended anode is eaten away to replace the loss and the solution suffers no great diminution of strength. The anode for copper plating consists of a piece of sheet copper, while for gilding, the anode should be of pure gold and can be made by melting gold foil scraps into a button and then hammering it into a thin sheet. All anodes should have perfectly clean surfaces in order that they may be readily acted upon.

PLATING.

With the battery in position and the jars filled to within about two inches of their tops with their respective liquids, the operation of plating is a very simple one. After the article to be plated has been made smooth and polished and properly cleansed in the cleansing solution, it is rinsed in water, and if other than steel, is immediately suspended in the plating solution from the copper rod "c." The corresponding anode is hung upon the other rod "d," when the deposit of metal at once begins. The length of time necessary to secure a good coating will vary somewhat with the strength of the solution, the intensity of the current and the metal to be plated. Usually from ten to twenty minutes will be sufficient, but a little practice will be necessary to determine the time and secure the best results. If on removal the deposit is found to be too light, the article can again be placed in the bath and more added. When finally removed from the bath it should be held in running water and then dried.

If the article has a dead finish when placed in the bath, it will present a similar appearance when plated. If polished in the first instance, the deposit will have a polished appearance provided the current be not too strong.

The process of plating steel differs from the one just described in requiring the article to be dipped for a moment in the sulphuric acid solution after it comes from the cleansing solution and before it is placed in the copper plating bath. After receiving a fair coating of copper it is washed and then placed in the gold bath and gilded as described.

The deposition of the metal in any case usually progresses more quickly and evenly when the article to be plated is separated by about two inches from the anode and is slightly agitated while in the bath.

When the battery is not in use the anode should be removed from the plating solution and the zincs be elevated above the liquid in the battery. Both jars should also be covered to protect them from dust.

www.ingramcontent.com/pod-product-compliance
Lightning Source LLC
Chambersburg PA
CBHW021808230426
43669CB00008B/675